MW00327143

YALE AGRARIAN STUDIES SERIES

JAMES C. SCOTT, Series Editor

The Agrarian Studies Series at Yale University Press seeks to publish outstanding and original interdisciplinary work on agriculture and rural society—for any period, in any location. Works of daring that question existing paradigms and fill abstract categories with the lived experience of rural people are especially encouraged.

JAMES C. SCOTT, *SERIES EDITOR*

James C. Scott, *Seeing Like a State: How Certain Schemes to Improve the Human Condition Have Failed*

Steve Striffler, *Chicken: The Dangerous Transformation of America's Favorite Food*

James C. Scott, *The Art of Not Being Governed: An Anarchist History of Upland Southeast Asia*

Edwin C. Hagenstein, Sara M. Gregg, and Brian Donahue, eds., *American Georgics: Writings on Farming, Culture, and the Land*

Timothy Pachirat, *Every Twelve Seconds: Industrialized Slaughter and the Politics of Sight*

Andrew Sluyter, *Black Ranching Frontiers: African Cattle Herders of the Atlantic World, 1500–1900*

Brian Gareau, *From Precaution to Profit: Contemporary Challenges to Environmental Protection in the Montreal Protocol*

Kuntala Lahiri-Dutt and Gopa Samanta, *Dancing with the River: People and Life on the Chars of South Asia*

Alon Tal, *All the Trees of the Forest: Israel's Woodlands from the Bible to the Present*

Felix Wemheuer, *Famine Politics in Maoist China and the Soviet Union*

Jenny Leigh Smith, *Works in Progress: Plans and Realities on Soviet Farms, 1930–1963*

Graeme Auld, *Constructing Private Governance: The Rise and Evolution of Forest, Coffee, and Fisheries Certification*

Jess Gilbert, *Planning Democracy: Agrarian Intellectuals and the Intended New Deal*

Jessica Barnes and Michael R. Dove, eds., *Climate Cultures: Anthropological Perspectives on Climate Change*

Shafqat Hussain, *Remoteness and Modernity: Transformation and Continuity in Northern Pakistan*

Edward Dallam Melillo, *Strangers on Familiar Soil: Rediscovering the Chile-California Connection*

Devra I. Jarvis, Toby Hodgkin, Anthony H. D. Brown, John Tuxill, Isabel López Noriega, Melinda Smale, and Bhuwon Sthapit, *Crop Genetic Diversity in the Field and on the Farm: Principles and Applications in Research Practices*

Nancy J. Jacobs, *Birders of Africa: History of a Network*

Catherine A. Corson, *Corridors of Power: The Politics of U.S. Environmental Aid to Madagascar*

Kathryn M. de Luna, *Collecting Food, Cultivating People: Subsistence and Society in Central Africa*

Connor J. Fitzmaurice and Brian J. Gareau, *Organic Futures: Struggling for Sustainability on the Small Farm*

For a complete list of titles in the Yale Agrarian Studies Series, visit yalebooks.com/agrarian.

CONNOR J. FITZMAURICE AND BRIAN J. GAREAU

Organic Futures

*Struggling for Sustainability on
the Small Farm*

Yale UNIVERSITY PRESS

NEW HAVEN AND LONDON

Published with assistance from the
Louis Stern Memorial Fund.

Yale University Press books may be purchased in quantity
for educational, business, or promotional use. For
information, please e-mail sales.press@yale.edu (U.S. office)
or sales@yaleup.co.uk (U.K. office).

Set in Janson type by Westchester Publishing Group.
Printed in the United States of America.

Library of Congress Control Number: 2016936482
ISBN: 978-0-300-19945-1 (hardcover : alk. paper)

A catalogue record for this book is available from the
British Library.

This paper meets the requirements of ANSI/NISO Z39.48-
1992 (Permanence of Paper).

10 9 8 7 6 5 4 3 2 1

To all the New England farmers working hard to find sustainable organic futures: thank you.

CONTENTS

PREFACE

The organic food consumed by the majority of Americans today has strikingly little in common with the organic food envisioned by farmers resisting the advancement of industrial agriculture in the 1920s, '30s, and '40s. The countercultural activists of the 1960s and '70s would have trouble recognizing the bulk of organic food we purchase today, as well as the bulk of farms where it is grown. What most folks consider organic today might even be unrecognizable to the organic consumer movement of the 1980s, which spent much time advocating for a safer food system. Since the 1990s, the very scale of the organic sector has grown to such an extent that organic produce is visible in virtually every grocery store in the United States.

The growth of organic farming has been nothing short of re-markable. Organic farming, once the ostensible stuff of Luddites, iconoclasts, tree huggers, and hippies (a stereotypical image still prevalent among students in our universities, at the very least!), is now a formidable mainstream feature of America's agricultural system. Such remarkable growth has also brought equally remarkable change. As organic farming has transitioned from a marginalized set of alternative farming practices to a federally recognized niche market within the agricultural mainstream, scholars and food activists alike have argued that the ecological and social ideals of the movement have largely given way to economic rationality and pesticide avoidance—at least in the corporate form of organic. Organic farming was originally intended to be smaller, agro-ecological (i.e., harmonizing the agricultural landscape with its surroundings), community-based and community building. Many would argue that contemporary organic farming is now merely an agro-industry that is averse to using chemicals, but very often folks continue to perceive it as a "movement."

Many observers have lauded the industrialization of organic agriculture for its ability to shift countless acres of farmland toward organic production, preventing thousands of pounds of petrochemical fertilizers, pesticides, and herbicides from entering the environment. On the other hand, critics have decried the corrupting influence of agribusiness in the organic movement, citing the watering down of organic standards and the loss of the community values originally associated with the movement. A third, newer group recognizes that this split between "for" and "against" organic is too simplistic and that some farmers have found a way to farm organically in a robust way. Although expressed in many ways, chief among these recognitions is the concept of "bifurcation," the observation that there are increasingly two organic sectors, one made up of relatively large farms that look more and more like the highly mechanized and highly capitalized conventional farms of agro-industry, and the other made up of small farms that are less mechanized, less highly capitalized, more likely to sell directly to the consumer, and (at least in some cases) less likely to consider profit ahead of other concerns (cf. Constance, Choi, and Lyke-Ho-Gland 2008).

While the praises and criticisms of contemporary organic agriculture certainly have their merits and offer important insights into the structure of modern organic agriculture, at the very least, the current system of organic agriculture presents us with an empirical puzzle. In an age of "industrial organic," what does "organic farming" mean to small-scale New England farmers operating outside the purview of agro-industrial organic? How are they, if at all, sticking to both new and age-old organic principles in an era that does little to support such determination? Where do these farmers see their organic products deriving their value—in the profit they bring or in something more complicated? And, critically, what is the lived experience of small farmers attempting to envisage an alternative agricultural reality amid the economic imperatives generated by an industrialized organic food system? How do these farmers contest the role of the market in the organic sector by enacting new market relationships and practices?

To answer these questions, we hope to extend and complicate the concept of bifurcation by paying attention to the relational, emotional, and moral underpinnings of organic farmers' market relationships. Outside agro-food studies, economic sociologists have increasingly pointed to the critical importance of such social forces in facilitating market exchanges and processes. Central to this work is the recognition that people often must match economic concerns and transactions to both their lifestyles and their meaningful personal, ethical, and moral relationships. Social forces do not merely constrain market processes; they actively construct them. Within the study of the contemporary agricultural economy, theories like bifurcation help explain the structure of the contemporary organic sector. However, the focus remains on the relationships found in agriculture's political economy, leaving the power of social relationships, moral obligations, and sentiment rather unquestioned in the shifting terrain of modern organic farming. At the same time, many of the now classic cases of the new, relational economic sociology focus on showing the ways market relationships and market logics enter previously non-market exchanges without terrible consequence, thanks to the careful matches actors forge. Here, however, we have a case where the entrance of the market into the organic sector is widely perceived as problematic. And yet small farmers appeal to new market relationships and market logics in their efforts to produce organic alternatives. As such, we hope to show that small-scale farmers actively make choices to farm organically that make sense to them in the face of various market forces—not solely because of them.

To understand how small farmers make good organic matches, we seek to revisit what organic farming once represented in the past, what it has become in the modern era, and what it has the potential to be in the future. Most important, this book is an attempt to understand what "organic agriculture" within the modern context means to people attempting to make a living as non-conventionalized, small-scale organic farmers in the existing farm system. On the whole, much of organic agriculture's potential to produce sustainability in our food system has been lost in the shuffle as organic farming has entered the agricultural mainstream; the values of agro-ecology, economic sustainability, and community have been replaced by the simple absence of (most) chemicals on many organic farms. We hope to provide a glimpse of how small-scale farmers make sense of this loss, as well as how they make sense of their own farming practices.

Certainly other benefits may have been gained by the expansion of organic farming into a formidable part of America's agricultural mainstream. However, the reality of the organic market is such that many small-scale farmers are not afforded the opportunity to set the rules of their own game. Yet they keep playing, struggling to remain alternative in light of obstacles that seem insurmountable. We want to contribute to the organic story by showing how some small farmers in New England are avoiding conventional organic. Their struggle is not unique. Small farmers around the country are finding ways to put ecology and community above profit. But the New England landscape is rather understudied in this regard, and the conditions for small-scale farmers are different here. Thus on the one hand, we try to show comparisons across regions by including the voices of farmers throughout the United States. On the other hand, we focus centrally on the lived experiences of those working on a particular small organic farm in New England—Scenic View Farm—because this method is the only way to truly tease out and understand the day-to-day practices, relationships, and commitments of farmers struggling to remain non-conventional. We show that Scenic View is not alone in its efforts in New England, and we find it fascinating that its farmers are redefining organic on their own terms in order to survive. (All names, places, and other identifying details of the case study have been altered to protect the identities of our research participants.) If there is any hope for non-conventional organic to remain vibrant and growing in the future, then these are the places it will happen. Here's to an organic future.

This story could not have been told without the help of many caring people. We are grateful for the insightful comments and suggestions made on many iterations of this project, including those of Paul

Gray and Julie Schor, both professors at Boston College. The Boston College Environmental Sociology Working Group, funded by the Boston College Institute for the Liberal Arts, gave us the chance to test out many of our ideas in a sociable setting. Boston College's Office of the Vice Provost for Research, the Boston University Department of Sociology, and the Office of the Dean of the Graduate School of Arts and Sciences at Boston University generously supported the completion of this manuscript as well. We are also grateful to Ashley Mears, professor at Boston University, and Kimberly Hoang, professor at the University of Chicago, for pushing us to consider the economic dimensions of farmers' lives in new ways. Emily Barman and Japonica Brown-Saracino, also professors at Boston University, provided much-appreciated advice and encouragement along the way. At Yale University Press, Jean Thomson Black provided excellent guidance and feedback (every author should be so lucky as to have Jean as an editor). Also at Yale, we thank Samantha Ostrowski for her help with keeping our manuscript in order (not an easy task with two authors at different universities) and our production editor, Ann-Marie Imbornoni. A very special thanks goes to Bojana Ristich, who was once again meticulous and exemplary in copyediting a book manuscript for Brian. Also, we are deeply grateful for the time and effort that the anonymous reviewers put into our project. These reviewers provided exceptional advice and were truly interested in making our book a better contribution to the field. None of this would have been possible without the farmers and farm workers who generously gave their time and energy to speak with us. Of course, we are especially grateful to those at Scenic View Farm and the other farms we encountered for opening their lives to our research. Thank you for welcoming us into your homes and fields.

Finally, on a personal level, Connor would like to thank Brian Gareau for his tireless efforts as a coauthor and dedication to making this book a reality. Connor would also like to thank his friends and colleagues, especially Claire Duggan, Cati Connell, Eric Buhr, and Carl Hudiberg, for providing listening ears—and lots of laughs—along the way. Connor is also grateful to Kerry and Dan, his parents, and Betty and Joe, his grandparents, for providing so much love, support, and encouragement throughout this process. Last but not least, Connor wishes to thank his partner, Bryan Kaufman, for sharing in all of the trips for barbeque brisket that made writing this book more manageable. Brian wishes to thank Tara, his wife, and his loving children, Delphine, Beatrix, and Leonel.

INTRODUCTION: CONVENTIONALIZATION, BIFURCATION, AND SOCIAL RELATIONSHIPS ON THE SMALL ORGANIC FARM

On a late summer evening in New England, an urban Whole Foods Market is buzzing with the frenetic energy of eager shoppers just off the clock. The scene can be a lot to take in, almost too much at times. Carts weave in and out of the aisles of produce in a delicately choreographed ballet, so tenuous, in fact, that the low buzz of shopping cart wheels ticking across the floor is very often punctuated by apologetic voices saying, "Excuse me," and far less apologetic guttural sounds, sighs, and moans.

In the midst of this frenzy, piles of glistening produce stand like so many totemic embodiments of the New England harvest. Men and women in smart suits and college students still carrying backpacks encircle mountains of fresh corn and shuck busily. Low wooden bins that resemble shipping crates contain a dazzling panoply of summer squash in seemingly every shape, color, and size. Farmers are present this evening too, their faces reproduced on signs that speak to the provenance of this seasonal bounty. There is "organic" produce in the market today, and every day, and plenty of "conventional" produce too. However, shoppers here must rely upon their own knowledge of these labels, and perhaps on the small blurbs beneath each farmer's face, to understand the meanings of these blanket terms. Achieving a more precise understanding of "organic" versus "conventional" is a hard task for shoppers in a supermarket. As the wheels of each shopping cart tick by the unmoving farmers' faces, what these terms mean to each farmer, in each farmer's fields, is impossible to discern.

A few days later on a sunny afternoon a farmers' market fills a downtown urban plaza with twenty or so vendors from local farms, bakeries, and greenhouses. Beneath the summer sun heirloom tomatoes arrayed in what seems to be a cheerful profligacy of colors glisten, still damp from the morning dew that enveloped them when picked a few hours earlier on farms some distance away. The pace is much slower here than during the after-work rush at the supermarket, although a person hurrying to catch a bus might occasionally shimmy to the front of a stall to pay and run off before the bus hisses out its exhaust and departs the corner. A flock of pigeons is startled by the sound, alighting over a family complete with wandering children and a stroller in tow. The ambrosial nectar of fresh summer peaches drips down the faces of the contented children. These peaches were purchased from a "low spray" orchard located just thirty miles north and west of the city center. A young couple buys a large bunch of Thai basil from a stall with a banner displaying the organic seal from the U.S. Department of Agriculture (USDA). At another stall, a woman looks approvingly at a mountain of eggplant, so purple as to rival the dyes of the Phoenicians. The woman asks the man at the stall if the eggplants are organic and is assured by him, the actual farmer of this produce, that they are, in fact, "better than organic" since the USDA regulations are concerned only with what farmers put on their fields and not the care with which they farm them.

It is easy to see in the observations above that the farmers' market takes us closer to understanding the links among the farmers, their farms, and their produce than a walk down the Whole Foods produce aisles. But even at this level, as one strolls through the farmers' market, farmers and their farm signs can easily become a blur of faces and images lining the stalls, their stories left untold in the brief exchanges that take place between consumers and producers. Perhaps, over a season, a customer may pick a favorite vendor and develop a relationship with him or her. At the market described above, many of the vendors recognize their regular customers and take a few moments to chat between sales. However, in fleeting conversations about the quality of this season's potatoes or the relative merits of one beautifully garish variety of heirloom tomato over another, it can be difficult to comprehend what it means for one farmer to be "better than organic" and for another to display the organic label. It can be more difficult still to piece together how those meanings—along with a host of other social and moral relationships—ultimately shape the types of farming practices the farmers are willing to use and those they would do anything within their power to avoid.

Such meanings, relationships, and commitments, and the powerful ways they shape the everyday practices of organic farmers, became suddenly apparent in the red wooden barn of Scenic View Farm, the main focus of our study in this book. Scenic View Farm is a small certified-organic vegetable farm with six acres of fields under cultivation, located in a picturesque New England town about an hour and a half drive from Boston. John and his wife, Katie, the farm's principal operators, wanted to check how the potatoes were doing while we spoke. "Let's go down to the garden and take a look," John offered, putting on his wide-brimmed hat. Every time we spoke about John and Katie's land, we spoke about their "farm." However, every time they spoke about it, they chose to speak about their "garden."

The word "garden" suggests a planted plot of land, often extending out from the home, while "gardening" is often a leisure activity. Scenic View Farm is nothing like a garden. John and Katie cultivate their six acres of fields with five additional employees, two tractors, a greenhouse, and a barn complete with a commercial kitchen. Yet such a framing profoundly shapes not only how John and Katie view their work, but also what agricultural practices they view as acceptable for their business and how they comprehend their relationships with their hired hands and customers. For example, framing their farm as an extension of their home makes John and Katie skeptical of using chemical pesticides on their land—even those allowable under the USDA's organic regulations. Enjoying observing the farm's wildlife while sowing lettuce seedlings and eating peas straight off the vine in the midst of harvesting made them even more circumspect of any form of chemical use. In deciding how to handle an outbreak of disease—whether to make use of allowable chemicals, to engage in other practices, or simply to endure an anxious period of watchful waiting—such understandings, commitments, and relationships played important roles alongside very real economic concerns about the success of their crops.

In the everyday social interactions, practices, and struggles of farmers to make their economic practices consonant with the meanings they attach to them, we can see the ways such forces profoundly shape what contemporary organic farming looks like in practice. In settings stripped of these crucial relationships, interactions, and personal meanings—such as the produce aisles in Whole Foods or even the stalls of the farmers' markets—it can be much harder to understand how the regulations and market conditions of the organic sector find expression in everyday farm tasks.

Understanding the Organic Farm: Conventionalization versus Bifurcation

Academic theories of organic agriculture have often run into similar dilemmas. As we will see in this book, organic agriculture has gone through significant changes since emerging from its social movement roots of the 1960s. It has grown as well as changed. Some would argue that nowadays much of what gets purchased under the organic label is produced under conditions nearly indistinguishable from its conventional counterpart. An extensive literature on organic has theorized these changes by looking at the political economy of organic farming (e.g., Guthman 1998, 2004a, 2004c). However, often lost in the shuffle of this approach are the complex personal and social negotiations in which farmers' decisions and practices are wrapped up on a daily basis. We hope to shift the conversation about organic agriculture in a way that will allow us to take seriously farmers' agency while still recognizing the economic forces that constrain them from using "perfect" organic farming practices. In other words, we strive to emphasize the profound importance of extra-economic considerations in shaping how small-scale farmers seek to forge an organic future in an economic world that is in many ways set against them.

The idea that contemporary organic agriculture has come to increasingly resemble conventional agro-industry and—perhaps more important—that the entry of agro-industry into organic farming has altered the playing field for all organic farmers has been expressed in the conventionalization thesis. The literature on conventionalization finds that organic standards have been loosened, allowing large-scale corporate farms to outcompete smaller growers without needing to concern themselves with agro-ecological commitments to building the soil or historical concerns for deeper ecological, social, and economic sustainability. The conventionalization approach was introduced to agro-food studies by researchers such as Daniel Buck, Christina Getz, and Julie Guthman in the late 1990s (Buck, Getz, and Guthman 1997) and has been championed in much of Guthman's work (cf. Guthman 1998, 2004a, 2004c). Since its introduction, the conventionalization thesis and its analytical approach, rooted in political economy, have become a dominant force shaping an understanding of organic farms and farmers.

The conventionalization thesis concentrates on a surprising and unfortunate paradox in the organic industry. Regulations—often viewed as a primary means of protecting consumers by ensuring the integrity of products and services—led to the watering down of organic standards. Due to the National Organic Program (NOP) of the USDA and its regulatory focus on substituting synthetic chemical inputs with organic alternatives—through which many chemicals *remain permissible* in organic farming—agro-industrial farms are able to merely swap out banned synthetic chemicals for allowable alternatives. This focus has left industrial farming practices writ large rather unchallenged, even on farms that adopt an organic farming method (Guthman 2004c). Central to this approach is the recognition that neoliberal forms of governance have shaped regulations in agriculture in such a way as to minimize the potential for organic to become truly ecologically sustainable and socially transformative.

Numerous scholars have applied the concept of neoliberalism to agriculture and specifically to alternative agriculture movements (cf. P. Allen 2004; Gareau 2008, 2013; Gareau and Borrego 2012; Guthman 2008b; McMichael 2010). Neoliberalism is commonly defined as a practical political economic project involving—to varying degrees of success—"the privatization of public resources and spaces, minimization of labor costs, reductions of public expenditures, the elimination of regulations seen as unfriendly to business, and the displacement of governance responsibilities away from the nation-state" (Guthman 2008b, 1172). Pressures on organic farmers come from changes in international and national regulations of the agricultural sector that open up the agricultural industry to more, and bigger, players.

Consistent with the work of food regime scholars (e.g., Buttel 2001; Mascarenhas and Busch 2006; McMichael 2005; Pechlaner and Otero 2008), as well as the perspective of Karl Polanyi (1944), we recognize neoliberalism in agriculture as a form of "neo-regulation" and not deregulation as the term readily implies. As the persistence of U.S. agricultural subsidies for large-scale farmers makes all too obvious, protections exist for the largest growers even at a time when such protections are ostensibly out of favor. In the "actually-existing neoliberalism" of contemporary agriculture, principles of deregulation have been applied unevenly (Brenner and Theodore 2002). Rather, certain types of regulations have proved critically important in producing neoliberal markets. While deregulation helps to open up markets and create demand, without rules to guide exchanges market actors often are cautious in making further investments (Fligstein 2005). In the neoliberal era, large corporate actors have disproportionate power to shape necessary market regulations, in part because they can always turn to "systems of private governance to settle contract disputes or produce common rules of property rights, governance, or rules [of] exchange" if favorable regulations seem unachievable (200). As a result of the alignment of corporate and state interests, the interests of "consumers, labor, or environmentalist groups" rarely make the regulatory agenda (ibid.).

In the agricultural sector, the state has operated as the facilitator of a new era of regulations that permit the introduction of chemicals and genetically engineered crops into the food system and make increasingly intensive corporate farming operations possible (Goodman, Sorj, and Wilkenson 1987; Busch, Burkhardt, and Lacy 1992). These regulations put financial pressure on small-scale famers world-wide and threaten the integrity of organic farming practices and markets (Pechlaner and Otero 2008; McMichael 2005). As we will see in the following chapters, organic standards in the United States have undergone a neo-regulation of their own, allowing for the entry of corporate actors with large-scale industrial farms into the organic sector and making it increasingly difficult for small-scale organic farms to remain economically viable while maintaining agroecological practices.

Guthman (2004c) identifies three main factors pushing organic farmers to shift to practices that look increasingly similar to those of their conventional counterparts. These include the focus on inputs instead of processes in the regulations, the ability of large farms to benefit from economies of scale and depress prices, and the tendency for farmers to cut corners when facing declining profit margins. While Guthman's theory was developed in the context of California (and she acknowledges the exceptional character of California's agricultural economy), the mechanisms she identifies have market-wide repercussions. The depression of prices by large-scale farms in California, for example, affects many farmers throughout the market given the fact that California produce (conventional or otherwise) is sold in grocery stores from coast to coast. While the effects of conventionalization may be more pronounced in places like California, the effects of conventionalized, industrial organic can scarcely be considered from a strictly regional perspective in an era when the conventional grocery store is the most common place consumers make organic purchases (Dimitri and Greene 2002, iii).

The explosive growth of organic food production has meant that modern organic cultivation is occurring on an unprecedented scale. For many, this is an undeniable gain. More land is being cultivated without potentially harmful chemicals, consumers have gained new access to organic foods (and at lower prices), and the potential for further expansion of the market continues. Even critical scholars who question the sustainability of organic farming on an industrial scale have noted the potential for modern organic farming—even if it is conventionalized—to promote positive changes. For example, some scholars have suggested that even if the environmental gains of organic agriculture diminish with conventionalization, the ability of green consumption of organic products to promote greater environmental awareness and social activism could persist (cf. P. Allen and Kovach 2000). Despite such potential benefits, however, the reality remains that many of the tenets of organic farming's social movement roots have been abandoned, at least in contemporary organic agriculture's more industrialized forms.

Consequently, a major question for agro-food scholars has been whether conventionalization is inevitable for all contemporary organic farms given current market expansion, competition, and the resultant depression of prices. Many have argued that at the very least, the tone of the conventionalization thesis is overly economistic and deterministic (Coombes and Campbell 1998). Instead, these observers acknowledge that while some of contemporary organic production increasingly resembles conventional agro-industrial farming, not all organic farms have abandoned the tenets of the organic movement, nor have they succumbed to economic pressures to conventionalize their farming systems. Instead, organic agriculture has bifurcated.

Large-scale corporate organic farms exist alongside smaller farms that remain committed to more holistically organic practices (rather than relying on input substitution) and continue to market directly to consumers (see Constance, Choi, and Lyke-Ho-Gland 2008; Coombes and Campbell 1998). While some have suggested bifurcation is a part of conventionalization (Buck, Getz, and Guthman 1997), others argue that the model of conventionalization is hardly inevitable. Small-scale farmers are not on an inescapable march toward conventionalization (Campbell and Liepens 2001). Instead, industrial organic and the organic movement operate in separate spheres, fulfilling different market needs and satisfying separate consumer desires and demands. Most important, smaller organic farmers often rely on direct marketing to consumers or fill niche markets for specialty products that allow them to escape the main pressures of conventionalization (Constance, Choi, and Lyke-Ho-Gland 2008).

A Theory of Everyday Economic Practices
on the Organic Farm

While the conventionalization-versus-bifurcation debate is important to organic agro-food studies, this book endeavors to explore organic farming from a slightly different perspective. There are economic and legal relationships that govern organic agriculture, and research on conventionalization and bifurcation has described and continues to investigate the effects of these relationships. Like the proponents of the bifurcation thesis, we do not believe that organic farmers inhabit a world in which the economic bottom line trumps all and in which idealism is inevitably watered down by the logic of economic competition. We know, even if only from our own lives, that our ethical commitments, our values, our hopes, and our dreams often lead us to act contrarily to our rational economic self-interest. These extra-economic commitments can serve to thwart the "economic push" in an ostensibly inevitable direction. And yet our data show there is not a separate organic market in which an ideologically motivated remnant of farmers persists uninfluenced by the rise of industrial organic agriculture. The fact that organic products can be procured not just at farmers' markets or food coops but also at lower costs than ever before in the history of modern organic farming from the nearest big-box store provides a crucial context for all organic consumers and producers alike. These are real-world economic pressures that have shaped the organic movement and influenced the lifestyles of even the smallest organic farmers holding on to ecological principles.

However, despite these economic realities, the legal and market-based forces shaping organic farming practices—rooted in organic agriculture's location within a broader political economy—do not operate in isolation of other social phenomena. Understanding market and regulatory forces also depends on understanding how these forces are implemented in everyday economic practices (Swedberg 2003). Indeed, numerous social relations—including institutional contexts, norms, politics, symbols, and meanings—shape what actors believe to be economically or legally strategic (Edelman and Stryker 2005; Edelman, Uggen, and Erlander 1999; Macaulay 1963). Within the study of organic agriculture, the conventionalization thesis has been critiqued for an overly deterministic view of regulatory and economic forces that mistakenly describe the social relations of organic farming as being "nothing but market" relations, a characterization that has influenced many analyses of economic life (Zelizer 2010). The conventionalization thesis falls short, then, when it argues that the economic sphere operates separately from others spheres of social life. As Viviana Zelizer explains, "With separate spheres, we have the assumption that there are distinct arenas for rational economic activity and for personal relations, one a sphere of calculation and efficiency, the other a sphere of sentiment and solidarity" (2007, 1059).

However, market relations and market pressures are rarely irreconcilably at odds with other social values. Anthropologists have long recognized that economic exchange is rooted in networks of reciprocity, sentiment, and trust (Mauss 1954). The choices organic growers make about where they are situated in agriculture are shaped by where they are located in a web of interpersonal, cultural, and moral relationships, not merely economic ones.

On the other hand, while the bifurcation approach acknowledges that there are viable farms that have managed to stay outside of the industrial-organic mainstream, like the conventionalization thesis the focus of bifurcation approaches has tended to be on these farms' market relations. The success of non-conventionalized organic farms is explained through appeals to these farms' structural positions, which shield them from market pressures or make them undesirable for corporate acquisition.

As a result, bifurcation approaches are similarly critiqued for over-emphasizing political economy at the expense of extra-economic factors, such as how farmers negotiate claims of worth (Rosin and Campbell 2009). Indeed, our findings point to the important roles that social and ethical considerations play in organic farmers' decisions to stay in small-scale agriculture. As a result, the conventionalization and bifurcation approaches are both important but partial accounts of the contemporary organic market. Just as the organic market is governed by economic and legal relationships, it is also shaped by equally significant *social* and *moral* relationships. The ways these other types of relationships actively shape organic farming practices has received far less attention in the literature.

The organic farming sector is a perfect site to explore how social and moral relationships shape the practices of economic life. Organic farms are dense sites of exchange relationships that depend upon how a farm is socially and economically situated. Moreover, given the ideologically driven social movement roots of organic farming, organic farms are also sites where people's moral values come into play in economic decisions and where social and economic relationships often overlap. Moreover, organic farming represents a unique case in that there is robust evidence that it is increasingly difficult for farms to remain alternative—whether through competitive pressures in the organic market or through the watering down of organic farming's meanings through permissive federal regulations (Buck, Getz, and Guthman 1997; Guthman 1998, 2004a). As a result, organic agriculture represents an important site to answer questions about how social and moral relationships shape economic practices and imbue them with meaning in settings where market forces might otherwise undermine them.

In response to the limitations of understanding organic strictly through political economy, some scholars have sought to use a different theoretical language to explain how farmers arrive at distinct sets of practices. The use of conventions theory, a sociological perspective focusing on how people coordinate the standards by which social worth is evaluated in exchanges (Rosin and Campbell 2009), is one example. Many scholars of organic agriculture, however, have called for greater acknowledgement that markets are *socially embedded systems*, emerging out of particular contexts (see DuPuis and Gillon 2009; Hinrichs 2000). Drawing from the pioneering work of Mark Granovetter (1985), such scholarship has called our attention to the social and political actions that constitute alternative systems of agriculture and the micro-politics of collaboration and networking from which new systems emerge. DuPuis and Gillon argue that "we need to ask the empirical and micro-political questions about alterity-creation [the creation of alternative modes of interaction] as an everyday process—as a process that reproduces itself on a day-to-day basis" (2009, 44).

Recent developments in economic sociology have called for moving beyond the view that markets are simply "socially embedded" toward a recognition that markets are socioculturally constituted systems of norms, values, and networks of interpersonal social relationships (Healy 2006; Zelizer 1988, 2010). Thus, attempts to understand local organic markets as embedded economic systems should move beyond "distinguishing more and less market-like transactions [by] recognizing that *every market* depends on continuously negotiated meaningful interpersonal relations" (Zelizer 2007, 1060). Markets are not just embedded in networks of social relations—they are in fact the products of social relationships. In this way, emotions and relationships, as well as socioeconomic realities, are taken seriously as foundations for decision making. Recognizing organic markets as socially embedded systems corrects only some of the deficiencies of the more economistic accounts since embeddedness approaches view social forces as impinging upon the otherwise "rational" and "asocial" economic sphere. Regardless of how rational some markets may seem, they are still socially constituted phenomena predicated on norms, values, emotions, and relations conducive to such practices.

Such scholarship has generated a new, relational approach to understanding economic exchange, an approach we believe can contribute important insights into how organic farmers understand their economic lives through meaningful social relationships. Insights gained from recognizing the market as socially embedded have paradoxically led economic sociologists to "take the market itself for granted" (Krippner 2001, 776). Instead, we need to take seriously the content of the social ties that make up markets, a view where the economic activity of markets is forged through "political contestation over shifting cultural beliefs that provide a template for interaction in market settings" (801). Rather than our standard view of markets as systems built up through social ties, markets and economic life are reframed as achievements that must be reached through interaction (Bandelj 2012, 177). "Relational work" becomes a mechanism through which economic outcomes can be understood.

Relational work in economic life can be understood as an intentional social action undertaken to facilitate economic outcomes or exchanges—often shaped by power asymmetries between participants. Relational work is important in accomplishing economic outcomes since "what is exchanged in economic transactions is not only financial resources in capital markets or human resources in labor markets but also emotional resources" (Bandelj 2012, 181). As a result, emotional resources, meanings, and practices of relationship management central to economic exchange are considered seriously. It is precisely these types of social relationships that have received less attention in explaining the shifting practices of contemporary organic farmers.

Our Approach to Understanding New England Organic Farming

In this book, we bring the concepts of relational work, emotion amid economic transaction, and "good matches" from economic sociology into conversation with the debates about the conventionalization and bifurcation of contemporary organic farming. In doing so, we provide a language that complements conventions theory for understanding organic farmers that need not appeal to dualistic binaries of either conventionalized or small-scale, corporate or movement, market driven or ideological. While studies in conventions theory have been applied to the multiple institutionalized standards of evaluation within the organic sector (cf. Rosin and Campbell 2009), our focus is on how organic farmers navigate personally meaningful social and economic relationships—including the affective, cognitive, and behavioral components of those relationships (Bandelj 2012) and the meanings they attach to their farming practices. These concepts, as we shall see, help us understand the relational work that allows small-scale organic farming to make sense in the lives and livelihoods of New England farmers. Here we are exposing more than just an economic niche within a bifurcated market; we are illustrating the makeup of these farmers' lives, of which the economy is but one facet among many.

Our "economic lives" are not a series of either/or decisions between love and money, and neither are the lives of organic farmers. Economic and emotional considerations are always mixed, and we constantly negotiate good matches between the two that, at the very least, allow us to get on with the economic activity of daily life while sustaining important relationships and the meanings we attach to our work (Zelizer 2007). The lives and practices of organic farmers cannot be divided into a world of more or less market-like relations. However, such mixings often produce moral, economic, and relational ambiguities. "Good matches" are a specific type of relational work that manages these ambiguities (Zelizer 2010).

Individuals work to match economic activities with social relationships, sentimental attachments, and understandings of what is appropriate. They do so through carefully negotiating the meanings, emotions, and practices associated with relationships of economic exchange (Zelizer 2012). From a relational approach to economic life, such matches are critical in this "reciprocal process" because "any misalignment in expectations and differences in interpretations between the participants exacerbates uncertainty and ambiguity" (Bandelj 2012, 194). From this theoretical perspective, small-scale organic farmers need not be either on the path to conventionalization or occupying insular refuges in a bifurcated organic market. Instead, they are seen as both: as farmers making decisions that attempt to achieve a workable match among the seemingly contradictory pulls from their economic concerns, their ethical commitments, their ecological values, their social obligations, and their deeply held lifestyle choices.

The concept of "good matches" has been employed to critically evaluate how individuals maintain a host of market relations where a conventional market calculus is commonly viewed as undermining ideological commitments, cultural values, or social relationships. For example, in the development of the U.S. organ donation system there was a need for appropriate matches between cultural beliefs and a pressing need for organs in medical markets. Organ procurement is a fraught terrain given cultural beliefs about the dignity of human life and the body. Yet with medical advances a need for organs arose. The current organ procurement system arose through the work of donation advocates—achieving a match between the market need to make organs available and cultural norms (Healy 2006). The form that organ exchange systems came to take was predicated on the matches donation advocates were able to forge as they sought to "incorporate donation into death rituals," frame donor enrollment as a morally virtuous decision, and imbue donation as a "kind of social immortality" (41).

While this type of match resulted in an institutionalized system for the appropriate transfer of human organs, other applications of good matches have taken the level of interpersonal interaction as their theoretical concern. The ways gallery owners sell works of art to investors, for example, is intimately tied to ideological beliefs that money corrupts the very purpose of art (Coslor 2010). However, the common sentiment that art exists for its own sake does not mean that gallery owners must either refuse to sell to investors or "sell out" and become profiteers. Instead, gallery dealers try to find "correct buyers" (buyers who would enhance the artists' reputations and appropriately handle works of art) and avoid traditional sales approaches like auctions—particularly with new artists—to properly match ideological beliefs about the nature of art with the need for profit (219). Good matches do not just happen; as these examples show, facilitating ideologically acceptable and culturally appropriate forms of economic exchange very often requires significant cultural and interpersonal negotiations.

While Zelizer's concept of the good match has the potential to return agency to market actors by providing a basis for understanding how market relationships are socially negotiated, every individual does not have an equal chance to negotiate favorable social, economic, and cultural exchanges. Rather, social categories of inequality, difference, and exclusion provide critical context for the types of matches an individual can negotiate; these matches can reproduce, reify, or, occasionally, erode the unequal positions of actors. Among Vietnamese sex workers, for example, the race and class backgrounds of both workers and clients profoundly influence the nature of their personal and business relationships. Sex workers of a lower socioeconomic status provide strictly sexual services for men of similar backgrounds since they lack the connections and cultural knowledge to navigate bars catering to Westerners and elites. Sex work looks very different for other women, whose cultural and economic resources allow them to circulate in social settings where they can negotiate long-term sexual and emotional relationships in exchange for remittances from clients from abroad and, in rare but significant cases, serious romantic relationships resulting in immigration and marriage (Hoang 2011, 2015).

In contemporary debates about organic agriculture, thinking about good matches and unpacking their relational underpinnings through ethnographic methods allows us to understand how small farmers balance the undeniable changes in the industry with their own farming aspirations without requiring a reductionist appeal to either the organic "market" or the organic "movement" as the sole, or even primary, determinant of their practices. As in other cases where the concept of good matches has been productively applied, we find that farmers work hard to balance the economic, social, and culturally meaningful content of their relationships. At times, such efforts may mean that organic farming decisions seem contradictory on the surface. (As we will see, some of our farmers choose to sell some produce at a break-even price, not out of stupidity or poor business acumen, but because this practice fits their values and matches the local network in which they are embedded and feel comfortable.) In other settings, where economic activity, sentimental attachments, and ideological commitments intersect and the concept of good matches has proven theoretically useful—for example, between married couples (see Zelizer 2005 for a discussion) or in markets for sexual and romantic services (Hoang 2015)—the social costs and cultural negotiations involved in navigating a good match are far more fraught than whether a farmer chooses to use cover crops or chemical inputs on his or her fields. As a result, we do not believe the seeming "messiness" of our account is a liability. Balancing sustainability and economic viability on a small farm is messy business—especially in an agricultural context that does little to support such efforts. However, by understanding how these farmers navigate their messy worlds, we hope to show the possibility of meaningfully sustainable farming practices in an era of agro-industrial organic food production.

Given this theoretical agenda, we chose to examine the decisions and practices of small-scale organic farmers on the ground, where good matches are made and enacted. First and foremost, this book is an ethnographic case study of a single organic farm—Scenic View Farm—based on participant observation in the farm fields, where good matches are made and carried out. In addition, we also include the voices of other small-scale farmers, through interviews with fifteen New England farmers and vignettes of farmers' stories from across the country, to get a deeper sense of the networks of social and moral relationships in farmers' lives beyond Scenic View. Our goal is not to present a regional case study. Instead, studying one farm closely—working alongside its farmers—allowed us to pay attention to the interactions, networks, and social relationships involved in good matching. (For a detailed description of our methods, see the appendix.)

In light of the pressures of the contemporary organic market, the everyday practices of the farmers with whom we worked and spoke needed to be carefully matched to their own deeply personal lifestyle commitments, the types of social and economic relationships they sought to cultivate on their farms and in their fields, and the meanings they attached to their work. Moreover, we show that these are matches that point to a potentially more sustainable future. In so doing, we hope to articulate a new way of describing contemporary organic farmers that pushes us away from thinking of them as trapped in a conventionalizing agrarian world that is overly determined by the economic. At the same time, we hope to provide a framework for understanding the real worlds of small-scale organic farmers who reject the approach of the organic industry and have become part of the "bifurcated" market, either by carving out outposts of resistance through meaningful decisions about their farms or by strategically occupying spaces that large-scale organic has yet to subsume.

In this way, this work follows in the footsteps of other studies of organic agriculture. Leslie Duram's (2005) work on organic farmers across the United States avoids the problems of many theories of organic agriculture by demonstrating how organic farmers—of varying scales—make organic farming work for them. Rural sociologist Michael Bell (2004), in his masterful regional case study of sustainable farming in Iowa, shifts the discussion to a "phenomenology of farming" to understand why some farmers do not adopt sustainable practices while others do. In many ways, Bell's approach implicitly moves beyond the fallacies of "nothing but the market" and "separate spheres" by examining the taken-for-granted assumptions farmers have about the work they do. These include assumptions not only about the agricultural market, but also about family, community, and ecology. As a result, Bell's account of sustainable agriculture is less about keeping pace with a conventionalizing industry—or keeping oneself out of one—than about a different way of knowing: a cultivation of a new vision of what farming is and could be. We too will examine the taken-for-granted assumptions of contemporary organic farmers—tacit knowledge about caring for the environment, their communities, their families, and themselves—but this time in New England.

By focusing on the practices of a single farm and drawing up a vocabulary for understanding economic action from outside the debates that have occupied the attention of observers of contemporary organic farming, we offer a new way of thinking about what is happening in the fields of today's organic farmers. The power of the case study is its ability to trace how general social forces shape local-level behaviors and outcomes (Walton 1992, 122). By focusing on a single farm, we are able to tease out a theoretical argument about how these forces produce particular farming practices that match farmers' lifestyles and commitments.

We show that the types of matches organic farmers can negotiate are shaped by the unique social, structural, and cultural positions of the farmers themselves. Most of the farmers we introduce in this book are privileged along the lines of race, class background, and cultural capital. As is common among famers in New England, most of our informants are white, many are privileged in terms of having family support or even family land, and the majority have college educations. As such, their experiences do not represent those of all small-scale organic producers in the United States. Other farmers' social positions do not afford them the same options and opportunities in negotiating good matches—for example, black farmers in the rural South or Latino farmers on the West Coast (cf. P. Allen 1993, 2004; Gilbert, Sharp, and Felin 2002; Pilgeram 2011; Wood and Gilbert 2000). However, our findings reveal a general need to theorize how all farmers interact with the political economy of the contemporary agricultural sector along with the social, cultural, and moral economies of their own personal relationships. Furthermore, while our sample is made up primarily of white farmers and farmworkers of privilege and considerable connections, these same spaces have been identified as sites of anti-racism discourse, solidarity, and institutional efforts to achieve social and environmental justice (Alkon and McCullen 2011). Understanding how and why such farmers choose and sustain these lifestyles could have wide-reaching effects.

Moreover, the types of good matches we observed in New England are important in that they represent the efforts of a group of individuals (shaped by and responding to their own particular context) to navigate the seemingly contradictory realities of contemporary organic agriculture in a way that they can take for granted, in a way that allows them to rest at night knowing they are doing what they think is best, and in a world that seldom rewards the small farmer (privileged or otherwise) over the conventional giant. By and large, the farmers we studied were able to forge viable matches between their values and their businesses. Empirically, understanding the relational work involved in forging such matches—and the types of farms and farmers able to negotiate such matches—is important for better understanding the contemporary organic sector. At the same time, understanding the conditions that are necessary for the types of sustainable practices we observed to represent viable options for farmers is also critical to expanding the reach of sustainable farming in the modern agricultural economy.

Overview of the Book

How has organic changed over the years? The first section of this book, "The Market," surveys the history of the field of organic production. In this section, we examine the transformation of organic agriculture from a fringe social movement to a mainstream industry, and we return to theories of conventionalization and bifurcation that explain this major shift through the political economy of agriculture. The section begins by "Making Sense of Organics," tracing the movement's early development to reveal what organic farming once meant to its practitioners and supporters. Next, in "Organic Hits the Mainstream," we examine the legislative processes that established organic farming as a stable market category—paving the way for industry to enter the field. The history of this transition is detailed and complex, involving the often-competing conceptualizations of organic held by multiple state, corporate, and social movement actors. Scholarship on the transition to federal organic regulations has pointed to the ways farmers' voices have been overwhelmingly excluded from the final regulatory process (see DuPuis and Gillon 2008). Certainly, organic farmers and movement NGOs both contributed to and resisted the ways organic agriculture was conceptualized in federal regulations, but we have chosen to focus our account on the historical forces leading to a conventionalization of the organic label, to which contemporary farmers must respond in their own ways. Finally, in "Why Supermarket Organic Matters," we make a case for the value that has been lost as organic food has shifted to become one option among many in the conventional food marketplace.

The second section of this book addresses "The Land." Throughout the first section, we see how economic rationality has become the driving force behind much of the modern organic industry and where a vision of organic farming motivated by multiple rationalities (such as environmental, social, and familial concerns) is more or less limited to small-scale farmers in a bifurcated market. As Michael Bell explains, "Farmers face the same basic issues of contemporary American life that we all do: downsizing, escalating economic competition, the pervasive sense of never having enough time to do anything, conflicts in the relationship of home to work, struggles to accommodate difference, loss of community, and resulting threats to self and identity. . . . But farmers face these issues in ways that are peculiar to their work and to the rural scene" (2004, 33). So, too, do contemporary small-scale organic farmers.

Confronted with the very real economic challenges presented in the first section, we consider how small-scale organic farmers make good, workable matches among these structural realities, the unique conditions of where they farm, and their own beliefs and values. Based on a case study of Scenic View Farm and supplemented with the interview data and accounts noted above, this section explores how organic farmers struggle to make sense of their daily practices within the larger economy and, ultimately, how they can manage to make them work. For small farms struggling to remain alternative, "the market" and "the land" are not mutually exclusive. These farmers inhabit both worlds and live according to both logics, as well as a host of others.

The second section begins with "A Sense of Place," situating our case study within the larger context of small-scale farming. Scenic View Farm is at once unique and highly typical. Numerous farms across New England, and countless farms across the United States, face the same challenges and engage in similar practices. "Amid the Chard" plunges us headlong into the daily lived realities of a small organic farm: pulling weeds and cultivating rows of Swiss chard. This chapter provides an account of one farm's struggles and triumphs in a growing season marked by miserably wet weather, insect infestations, and the looming threat of plant diseases. Next, in "Who Farms?" we will glimpse some of the deeply personal motivations that shape the diverse organic practices at Scenic View Farm. In the next chapter, "A Sea of Brown Bags and the Organic Label," we get a glimpse of the marketing practices of a small, organic farm. From mundane tasks like packing one hundred brown bags for the farm's Community Supported Agriculture program, or CSA, we will gain insight into the economic challenges faced by small-scale organic farms—from a farmer's earnest desire to enter the middle class to the challenges of competing with the industrial food system. Finally, in "No-Nonsense Organic," observations from Scenic View Farm will challenge us to think about the ways beliefs about the environment, health, and product quality become solidified in the everyday concerns and economic realities that make up the relationships of today's small-scale organic farmers. Throughout these chapters, we draw from our interviews with other New England farmers and the experiences of small farmers across the country to show that Scenic View Farm is not alone in its effort to make good matches. By drawing from other farmers' experiences, we hope to show the explanatory potential of the concept of "good matches" for understanding how small farmers opt out of a conventionalizing organic sector, while we stay close enough to the experiences of the farmers in our case study to explore how more sustainable matches can be made.

The conclusion of this book, "An Alternative Agriculture for Our Time," will lead us to consider Scenic View, and other farms similar to it, in the broader context of the contemporary organic market. Can such small-scale organic farming represent a sustainable vision of agriculture with the potential to be truly alternative to agro-industrialization? Our case study is a glimmer of hope, showing that all that once comprised the organic movement has not been lost in the pursuit of industrial-sized profits and supercenter-sized public profiles. However, it is just a glimmer. Remaining alternative is a constant, ever-evolving struggle for farms operating within a conventional market framework. Community Supported Agriculture often degrades into a novel form of consumption lacking any serious community support, and there is constant pressure for small farmers who have opted out to get back on the treadmill of work-intensify and spend. As such, we show that there are pressures that threaten the viability of small farmers' best efforts toward more sustainable lives and livelihoods within the organic farm system. If good matches require constant negotiation, we show the places where those negotiations are most fraught.

Here again, a relational approach to economic exchange becomes invaluable. This book is primarily about the production side of the organic sector, of the process by which organic produce gets from the ground to the table and all the steps in between. A relational approach to economic life, though, highlights how all economic actors in an exchange—including consumers—play critical roles in negotiating its meanings and the social, ethical, and economic commitments their exchange relationship requires. If "relational work is a process that targets relationships" by negotiating boundaries of inclusion and exclusion through associated practices (Bandelj 2012, 182), the concept includes not only relationships between alternative organic farmers and their land, but also the ties they forge with the alternative consumers who validate the economic work farmers undertake in their fields. Improving the opportunities for sustainable relationships on organic farms will require changes in the ways consumers relate to the agricultural system.

Alternative farming initiatives—from farmers' markets to CSA programs—are often predicated on relational networks of exchange. Juliet Schor and Craig Thompson, for example, find that alternative forms of economic exchange and alternative consumer lifestyles are "creating new ways for people to meet their needs by sidestepping malls, big-box stores, and professionalized services" (2014, 3). Instead, these alternative consumers desire to be more involved with the production of goods and services, either through establishing local connections in local economies or by becoming directly involved in production. A do-it-yourself ethos and a do-it-together focus on community are the hallmark of what they term "plentitude practitioners." "Face-to-face, personalized economic exchanges" are resurfacing as highly valued experiences, "not just for their economic viability but on their own terms" as consumers increasingly desire "to know the farmer who grows their food and provides the milk they drink" (14).

We can think about the consumers who support our main case study site, Scenic View Farm, as well as "plentitude practitioners" more generally, as engaging in the relational work of matching the right types of meanings, sentiments, and relationships with these new forms of personalized economic exchanges. Thompson and Coskuner-Balli explain: "[The local food movement's] alternative ideological frame, and its corresponding mode of communal consumption experiences, enables CSA consumers to perceive the unconventional demands and transaction costs imposed by this countervailing market system as socially redeeming benefits" (2007a, 137). As a growing number of consumers seek more sustainable, local, personalized economies, they often must match their lifestyles and values to a set of consumption practices that require far more time, physical and emotional energy, and commitment than a simple trip to the grocery store. Moreover, many of the relationships local farmers try to cultivate with their consumers depend upon such investments. Often, however, farmers have struggled to get the community support that the alternative food systems they envision require (see DeLind 1999, 2011). Imagining a more organic future, therefore, will necessitate considering the cultivation of new types of good matches— such as those being forged by "plentitude practitioners"—that can encourage community investment of more than mere capital in the farms and farmers struggling for a new agrarian reality. Facilitating and fostering these types of matches could represent our best chance at a more sustainable, more equitable, and more organic future.

PART I

The Market

one **MAKING SENSE OF ORGANICS:**
A BRIEF HISTORY

In 2001, just three years after John began Scenic View Farm, he chose to obtain certification under the newly implemented regulations of the National Organic Program of the USDA. That same year, ordering a drink at a ubiquitous Starbucks coffee shop was complicated by the added choice of organic milk. Perhaps readers can remember the first time they tried to order a drink while taking advantage of this novel option. The request for a "Tall, no whip, pumpkin spice latte with organic milk, please," would leave most people tongue-tied and fumbling for words. Granted, Starbucks no longer offers an organic milk option; it was discontinued in 2008, after the company decided to use only synthetic-growth-hormone-free milk in all of its drinks. But for seven years, in countless Starbucks locations across the country, Americans could consume organic dairy.

When Starbucks began offering an organic milk option and Scenic View received its organic certification, John was cultivating just two acres of land. While the practices of organic farms that can produce on a scale to supply a company like Starbucks are not directly comparable to those of farms like Scenic View, the juxtaposition presents a puzzle. Organic agriculture—once a fringe movement with countercultural associations—now looms large in the American dietary consciousness, with implications for both industrial food producers and small-scale farmers like John.

Some might argue that the organic milk offering and withdrawal merely demonstrates the limitations of organic food as a "yuppie" movement. Starbucks offered it, after all. However, the fact is that organic products are all around us. McDonald's offers organic coffee, Walmart has entered the organic food business in a big way, and Kraft has even formulated an organic version of its iconic "Blue Box" macaroni and cheese. Despite the increasingly visible presence of organic foods in our consumer culture, questions remain about what effect this explosive growth—and profitability—has had on the farmers who produce these foods. Even more fundamentally, for many people answering the age-old question of "What's for dinner?" the notion of "organic food" remains a nebulous concept with variable associations ranging from health food to sustainable food—and everything in between.

According to the USDA, "Organic agriculture is an ecological production management system that promotes and enhances biodiversity, biological cycles and soil biological activity. It is based on minimal use of off-farm inputs and on management practices that restore, maintain and enhance ecological harmony" (cited in Gold 2007). This is a very good definition of the philosophical heart of organic agriculture. However, the way this concept translates into actual regulations and practices has been a more challenging problem. Often, for the USDA organic food is simply food produced without the use of synthetic fertilizers, pesticides, herbicides, fungicides, or other chemical inputs. But we are getting ahead of ourselves. Organic farming was being discussed and defined long before the USDA decided to regulate it.

The story of the organic farming movement, and even the very definition of organic farming that we have inherited, is a story of people—people pushing back against the perceived encroachment of industrialization into their lives and onto their dinner plates. The organic movement's story is one of resistance. It is not a story of mere ideas or simply of government institutions like the USDA changing farming practices. Behind the development of organic farming are people with tangible concerns about their environment, their health, and their way of life. In order to understand what "organic" was and is and how it came to be the fastest-growing sector of the American food industry, we must first understand the people and the social contexts that fostered its development.

The earliest conceptions of organic agriculture as an alternative system of farming arose as some farmers began to challenge the unchecked advance of industrialized practices into their fields. They were a reaction to some of the most profound changes in the history of modern agriculture, the roots of which stretched back to the enclosure of the English countryside. In the eighteenth and nineteenth centuries, the enclosure movement in England was at its peak, transforming common grazing land into private property (Polanyi 1957). The result was of tremendous consequence. Landowners could farm their privately owned tracts intensively—increasingly via machines—for the first time. According to Peter Linebaugh in *The Magna Carta Manifesto*, enclosure and its menacing partner, slavery (which produced unpaid laborers for the new intensive-farming operations), "ushered industrial capitalism into the modern world" (2008, 94).

Of course, the consequences of enclosure were not unidirectional. Rather, agricultural historians have pointed to the ways in which enclosure—allowing for crises of overproduction and dramatic declines in prices in cereal grain markets, for example—led to the necessity of alternative forms of agriculture. Far from propelling farming inevitably toward a reliance on cereal grains, the low prices wrought by new surpluses led many farmers in Britain to grow novel crops such as woad—a botanical source of blue dye. As historian Joan Thirsk notes, "A single generalization which more accurately summarizes the trend . . . would underline the new flexibility introduced into land use. . . . Instead of retaining a relatively permanent distinction between arable and pasture, and a relatively strict arable rotation of winter grain, spring crops, and fallow, it became common practice to lay down old ploughland to lees (or temporary pastures) . . . and then plough again" (Thirsk 1997, 24). Nevertheless, the enclosure and privatization of agricultural land allowed for intensified production to feed burgeoning industrial populations, often at the expense of the land itself. "Intensively held land is indeed held as private property," and during the enclosure movement "smallholders and landless individuals all suffer . . . and the privatized property may be exposed to environmental degradation" (Netting 1993, 326). By privatizing ownership, enclosure spurred investment in intensified farming practices that led to the increasing rationalization of agricultural practices.

As agriculture "rationalized" and became less of a way of life and more of an economic pursuit, the study of agriculture as a "science" commenced. In the first half of the nineteenth century, with enclosure largely complete, scientists such as Sir Humphry Davy and Justus von Liebig began making strides in understanding plant nutrition. They realized that it was not manure that plants needed but the minerals the manure contained. In effect, they realized that "inorganic mineral fertilisers could replace manures and bring agriculture into the scientific fold" (Kristiansen, Taji, and Reganold 2006, 4). The depletion of soil fertility was a major problem in the 1850s, as Europe "scoured the globe for natural fertilizers" and went on to develop synthetic fertilizers to feed the growing population, first in Britain, then abroad (Foster 2002, 155–157). It was precisely this new scientific rationality that the first organic farmers were struggling to resist.

With every great advance of the economy, some individuals end up on the losing end. As industrialized agriculture advanced, such individuals often turned out to be the farmers. Organic agriculture began in the early twentieth century as a place-conscious, ecologically concerned, farmer-based response to the problems of industrialization occurring within the food system, particularly in England. While early critics of capitalism such as Karl Marx and, later, Karl Kautsky (1988) commented on the relation between soil infertility and an ever-expanding agricultural system that "prevents the return to the soil of its constituent elements consumed by man" (Marx 1977, 637), it was not until 1924 that the Austrian philosopher Rudolph Steiner offered one of the earliest articulations of an "organic" farming ideology as a social movement in a series of lectures on what he called "biodynamic farming" (Lockie et al. 2006, 7).

Steiner, the founder of anthroposophy, was a curious individual. Part philosopher, part occult minister, Steiner's philosophical ideas were so strange that his biographer, Colin Wilson, was reluctant to publish anything about him. Wilson was concerned that Steiner's claims "created such a sense of frustration 'that even the most open-minded reader' would soon 'give up in disgust'" (cited in Lachman

2007, 3). However, Steiner's complex, spiritual conception of the world was precisely what led him to view farms as living systems in need of understanding and nourishing rather than simply fertilizing (Lockie et al. 2006, 7). Such a view stood as a direct challenge to the new science of industrial agriculture.

While Steiner articulated a spiritual and philosophical challenge to farming methods that ignored viewing the agricultural system as a whole, the British agricultural scientist Sir Albert Howard laid the foundation for the modern organic food movement over the course of his career from the 1920s to the 1940s. Howard was a trained scientist aware of the revolutionary new chemical fertilizers, but "his upbringing on a Shropshire farm made him highly skeptical of the approach" (Kristiansen, Taji, and Reganold 2006, 4). Working at an agricultural experiment station in India, Howard observed local farming techniques and noticed that the health of farm animals was correlated to how the food they ate was grown (ibid.). These findings and his personal experience led him to question the wisdom of the new, rationalized agriculture.

Howard was resisting what he saw as a shortsighted approach to agriculture. In an attempt to transform modern farming into a scientific industry, farmers were feeding only their plants—not their soil. For Howard, the health of the soil ultimately led to the health of humanity, and the development of chemical fertilizers was undermining the health of agricultural systems (Fromartz 2006, 6–10). In *An Agricultural Testament*, published in 1940, Howard decried chemical fertilizers, arguing, "Artificial manures lead inevitably to artificial nutrition, artificial food, artificial animals, and finally to artificial men and women" (cited in Fromartz 2006, 9). Meanwhile, the failure of modern agriculture to care for the soil became all too apparent during the American Dust Bowl of the 1930s, and concerns about soil erosion led to the early application of ecological thought to agriculture, with many "prominent New Dealers committed to the idea and reality of a planned, permanent agriculture to avert chaos and ensure a future for American civilization" (Beeman and Pritchard 2001, 11).

Building on the intellectual work of these early proponents of ecological farming techniques, some farmers began to put these ideas into practice, challenging the notion that new industrial farming methods were improving the lives and outputs of farmers. In the late 1930s, J. I. Rodale established an agricultural experiment station in Emmaus, Pennsylvania, to develop and improve upon organic farming techniques. The connections to Sir Albert Howard were clear: in 1942, when Rodale began publishing the magazine *Organic Farming and Gardening*, he chose to list Howard as his consulting editor (Francis 2009, 7). The work of Rodale not only developed organic farming practices, but it also showcased those experiences in print.

In England, Lady Evelyn Balfour also began to apply the new organic ideas in the fields, starting a series of experiments in 1939 to compare industrial agriculture with the newly articulated organic methods. Lady Balfour, the daughter of landed gentry who supported her education, bought a farm in Suffolk that had fallen into disrepair. On this land, she began the first longitudinal study to compare the ways organic and conventional farming practices affected the health of the soil, plants, and livestock (Inhetveen 1998). Called the Haughley Experiments, Balfour's trials continued through the 1960s, profoundly shaping the emerging organic movement (Francis 2009, 6; Kristiansen, Taji, and Reganold 2006, 5).

Balfour's study was designed to be wholly scientific. "No loophole must be left," she wrote, "whereby any other explanation than that of the different soil treatments can be held to explain any differences that may result" from the field experiments (cited in Reed 2001, 138). However, Balfour—much like Rudolf Steiner—was deeply influenced by New Age spirituality, along with ideas from the emergent field of ecology. Indeed, Balfour titled her influential bestselling book *The Living Soil* and spoke of organic practices as fostering "life energy." As a result, Lady Balfour rejected the use of randomized plots (a method that still dominates agricultural science) as fragmentary and "incapable of throwing light on biological interdependencies in a functioning whole." Instead, she constructed a series of closed farming systems on her land so that "interdependences between soil, plant and animal, and also any cumulative effects could manifest" (Balfour 1977). These studies led to the founding of the Soil Association in 1946, an organization that continues to influence British organic farming to this day as the largest certifier for the United Kingdom's organic products (Soil Association Certification 2013). As Rodale, Balfour, and others began putting the early critiques of synthetic fertilizers into practice, they demonstrated that challenging the growing industrial system of farming was possible.

The early pioneers of the organic movement were, by and large, a conservative group. They were often involved in far-right political organizations, and the connections they saw between the health of the soil and the health of human beings were often tied to ideas of racial determinism and eugenics (cf. Lockie et al. 2006, 2). Their commitment to organic farming did not arise in a social vacuum. Rather, it arose out of a struggle against the unabated advance of industry and science into agricultural production. However, they were not merely clinging to anachronistic farming techniques in a changing world. They wanted progress, but they were wary of the direction "progress" was taking.

The Countercultural Movement in Organics

Although the commitment to healthy soils remained, the conservative bent of the organic movement began to change in the 1960s as farmers and concerned citizens started to resist a new advance: the increasingly visible problem of environmental degradation. While the organic movement may have its roots in the early responses to problems associated with the industrialization of food production, it flourished during the counterculture movement of the late 1960s and '70s. For example, between 1970 and 1980, the number of farmers' markets, an alternative to conventional retail venues, increased dramatically; "Markets in some states increased tenfold, with the national total rising nearly 500 percent by one estimate" (Brown 2001, 655).

Such growth marked a radical transformation of the organic move-ment, as the concerns of the "new" organic pioneers refocused those of their predecessors fighting the advance of chemical fertilizers.

By the end of the 1950s, it had become clear that the organic move-ment had "lost the post–World War Two argument over the direction of agriculture" (Kristiansen, Taji, and Reganold 2006, 6). Indus-trial agriculture had become the norm, benefiting from the rapid advances in chemical technology fueled by the war. "Scientists had discovered that nitrogen-rich chemicals used to make wartime ex-plosives could be repurposed as synthetic agricultural fertilizers" (Beavan 2009, 124). The "repurposing" of bomb constituents seemed like a win-win situation. Factories that would have become obsolete stayed open, and farmers got a steady supply of modern fertilizers. Organic farming practices seemed even more antiquated then they had been in the 1920s and '40s. After all, synthetic fertilizers were seen as the saviors of the American farm. As Barry Commoner wrote in *The Closing Circle*, "This new technology has been so successful . . . that it has become enshrined in a new kind of farm management so far removed from the ancient plan of farming as to merit a wholly new name—'agribusiness'" (1971, 147–148). As a result, the struggles against environmentally degrading farming practices that emerged in the 1960s gave the movement new life.

As the American public and the world became aware of the envi-ronmental havoc caused by chemicals such as DDT through Rachel Carson's *Silent Spring* (1962), an ecological approach to farming that emphasized the health of the whole agrarian system gained new cur-rency. Organic agriculture's focus on the health of the soil was no longer fueled solely by philosophical concerns about the use of "artificial" substances to improve fertility. Rather, individuals began to challenge and resist what such "artificial" substances were doing to the world around them.

As we look back on the organic movement from the relatively tame "Age of Whole Foods," it is difficult to imagine exactly how radical this period of resistance was. Within the counterculture movement, the burgeoning organic movement expanded its focus beyond the mere health of the soil and food. Rather, it fostered "not just an alternative mode of production (the chemical free farms), but an alternative system of distribution (the anti-capitalist food co-ops), and even an alternative mode of consumption (the 'countercuisine')" (Pollan 2006, 143). The organic activists of the 1960s saw the connections between the environmentally destructive practices on farms across the nation and the corporations with vested interests in unloading toxic chemicals. It was not enough to simply change farming practices; the corporatized industrial system of agriculture had to go.

Among those resisting an environmentally reckless corporate food system were the Diggers, who distributed free food and a message in the Haight-Ashbury section of San Francisco during the mid-1960s. Their goal was to "use food as a medium [to develop] collective social consciousness and social action" (Belasco 2007, 17). For the Diggers, food was "at the center of an activist program based on an emerging ecological consciousness" (18). Others joining the resistance in the organic movement were Robin Hood's Park Commission. In April 1969, this group of agricultural activists seized a vacant lot in Berkeley, California, planted vegetable seeds, and renamed it People's Park. The poet Gary Snyder called People's Park "a guerrilla strike on behalf of the 'non-negotiable demands of the earth'" (cited in Belasco 2007, 21). In the turmoil of the counterculture, food was always more than "food." The organic movement of the late 1960s and '70s brought an environmental and political agenda home for dinner.

Indeed, in the turmoil of the 1960s and '70s food became a central concern for a variety of important political struggles. For instance, the Black Panther Party (BPP) established a free breakfast program for schoolchildren out of a church in Oakland, California, in September 1968. Reflecting on the founding of the breakfast program, former BPP chairperson Elaine Brown remarked, "Because we are so used to the capitalist construct, it doesn't occur to us that we have a human right to eat; because if you don't eat you will die, it's not complicated. So, if there is a price tag to eating, then there is a price on your head, because the minute you don't have enough money to eat, you're slated for death" (cited in Heynen 2009, 411). The place of food in society was being challenged as a social justice issue, and access to healthy food was being reframed as a fundamental right. The BPP breakfast program would develop a far greater reach than the nutrition initiatives of the Diggers. Just a year after the first program was established, the program became a mandatory action for all forty-five BPP chapters nationwide. Thousands of schoolchildren eventually received nutritious food through this program, which put pressure on Congress to permanently authorize federal funding of free and reduced-price school breakfasts for America's children (Heynen 2009).

As increasing and ensuring access to healthy food shifted from a farmer-based act of resistance to a broad-based social movement over the course of the 1960s and '70s, figures from the organic movement's past were thrust into prominence. For example, J. I. Rodale gained an almost cult following. In 1971, the *New York Times* ran an article on Rodale titled "Guru of the Organic Cult." In it, Rodale commented on the radical transformation of organic farming into a social movement: "It's made me so much happier. In the old days, I used to get such a clobbering and insulting, you know, and if I wasn't so well nourished, it would have affected me, but I stood up under it, because I had plenty of vitamin B, which is the nerve vitamin" (cited in W. Greene 1971). For Rodale, resistance had paid off. Maybe it was the B vitamins that allowed him to persevere, but there was something much larger at work in society.

The 1970s were a time of radical social change, a period in which social movements of all sorts thrived. Indeed, the author of the *New York Times* piece on Rodale tapped into this sentiment, saying that "proponents of organic food talk, as one does in these times, of a movement; the organic movement" (W. Greene 1971). What were "these times"? Society was mobilized against war, imperialism, racism, capitalism, gender inequality, and environmental degradation.

Rodale's resilience had paid off, in large part because the organic farming movement of the 1920s, '30s, and '40s was swept into the general social rebellion of the larger countercultural movement. Society was on the move, mobilizing against the ills of "the system," and the organic movement was brought along for the ride.

Looking back, we can see why Rodale would be so much happier in 1971 than he had been thirty years earlier. The popular *Whole Earth Catalog*, a forum for advertising products with countercultural appeal, listed Rodale's magazine, *Organic Gardening*, as "the most subversive publication in the country" (W. Greene 1971). This is the same magazine Rodale began publishing under the name *Organic Farming and Gardening* in 1942. It took almost thirty years, but Rodale's periodic handbook on naturally growing vegetables came to be seen as revolutionary among a wider countercultural audience. Consequently, the mainstream establishment would soon put the organic movement under close scrutiny.

How can a garden guide become so subversive, dangerous even? Tomatoes, cucumbers, and eggplants are just food; but when they were grown without pesticides and herbicides and using natural processes, they became so much more. Rodale, who had plodded along through nearly thirty years of marginality in the farming community, was not so revolutionary. He had been pursuing the same goals for most of his life—seeking to nourish the health of humanity by nourishing the health of the soil. Certainly, he had a vision of changing the agricultural landscape of the United States; however, a visionary and a revolutionary are not coterminous.

What was truly revolutionary was the way in which the counterculture was able to mobilize existing movements to resist the perceived ills of the larger society. A tomato becomes revolutionary not by any magical process or natural fertilizing regime. Cow manure is no more intrinsically destabilizing for a social system than synthetic fertilizers. A tomato becomes revolutionary when it is reframed as such. When you vote with your shovel, fork, and spoon, your very act of eating and growing can challenge the status quo.

Political Backlash from the Mainstream

The fact that the countercultural movement had succeeded in making the personal food choices of American consumers political (to use the language of a popular countercultural colloquialism) did not usher organic agriculture into the agricultural mainstream. By the end of the 1970s, organic farming enjoyed greater popular support; however, organic farming practices were still viewed with considerable skepticism. Rodale and other organic supporters were still subject to intellectual clobbering and personal insults; they now just had more company. Some of these insults came from the agribusiness elite. For example, Henry J. Heinz II of the H. J. Heinz Company lambasted the burgeoning organic movement, saying that America was "a nation of nutritional illiterates. Food faddism advocates are persuading thousands to adopt foolish and costly eating habits" (cited in Jacobson 1972, 33). The attack levied by Heinz was remarkably similar to accusations of the organic movement against food industries; in the turmoil of the 1970s, the argument cut both ways.

The insults came in many forms. There were cultural critics of the movement. For example, a book reviewed in the *Chicago Tribune* in May 1974 disparaged organic consumers, whom it described as carrot-juice sipping "anarchlings," little anarchists who fear not only authority, but mass society in general (Petersen 1974). The book in question, *The B. S. Factor*, poked fun at the countercultural practice of labeling the naturalness of foods—it was "B. S." since even man-made foods were "natural." The reviewer praised the book since he now understood why organic food supporters were "such a bore" (Petersen 1974, 16).

The organic movement would brush off the attacks of the agribusiness elite and authors coining novel terms like "anarchling." The more significant challenges came from the federal government. Earl Butz, the U.S. secretary of agriculture under the Nixon administration, was a galvanizing figure in U.S. farm policy. In many ways, he was an advocate for American agriculture. Butz was quoted in the *New York Times* as reveling in having "got things turned around in the Farm Belt" (Duscha 1972). And in many ways he had. Food prices were up, and he defended the rise against strong criticism from the administration. Butz adamantly believed that farmers deserved more pay for what they grew.

In light of the criticisms leveled against him, Butz reminisced, "When President Nixon asked me to come to Washington, he told me he wanted a vigorous spokesman for agriculture. I told him at the time I was sworn in, 'Mr. President, you may have a more vigorous spokesman than you want'" (cited in Duscha 1972). However, Secretary Butz was a vigorous spokesperson for a particular type of agriculture: agriculture centered on commodity crops grown on a large scale—particularly corn. In fact, Butz—a trained agricultural economist—helped to popularize the now often maligned term "agribusiness" (ibid.). Secretary Butz left us with several additional phrases for which he is remembered: plant "fencerow to fencerow," "get big or get out," and "adapt or die" (ibid.; Pollan 2006, 52). For Secretary Butz, though, organic farming advocates were just plain foolish.

In August 1972, Secretary Butz gave his official estimation of organic farming. It would be a great concept, he said, "if we [could] just figure out which 50 million Americans we want to let starve" (cited in DeVault 2009; Jacobson 1972). The organic movement was growing, but still it remained on the fringes, well outside the American agricultural mainstream. Community gardens were planted in vacant lots, food coops sold organic food, and gardening magazines had become subversive. Organic was a movement, but for most Americans it remained a movement to be viewed with a fair amount of skepticism, if not outright derision. And for its supporters, organic farming remained a form of resistance.

Although concerns about environmental degradation gave the organic movement broader appeal, it remained on the fringes of American society. Few people were committed enough to abstract notions of environmentalism to change their behavior. Food is a deeply significant aspect of our culture. In many ways, our tastes are the embodiment of what we value. Environmentalism simply was not powerful enough to change those engrained values. It would take one more wave of resistance to bring the organic movement out of the counter-cultural terrain and into the center of America's dietary consciousness.

That final push came during the 1980s. While previous genera-tions had supported organic agriculture in order to resist the early industrialization of agriculture and the environmental havoc that industrialization caused, it was concern over food safety that finally brought organic agriculture into the national spotlight. The resultant growth in organic food production can only be described as "explo-sive" (Kristiansen, Taji, and Reganold 2006, 7). Concerned citizens supported organic agriculture as a way to resist the unsafe food pro-duced by the conventional food industry.

Throughout the 1980s and into the early '90s, wave after wave of panic swept the nation as food scares became a seemingly everyday occurrence. Many of the scares were simply tied to industrial food processing, with numerous cases of botulism making headlines. For example, in May 1981, thirty states were affected by a federal warn-ing not to eat canned mushrooms tainted with botulin toxin (Asso-ciated Press 1981). A year later, 172 people in the southern United States were stricken with yersiniosis—a bacterial infection causing extreme abdominal pain—from milk processed in a Memphis plant. It was the nation's largest outbreak of the infection (Associated Press 1982). While certainly alarming, food poisoning related to food pro-cessing was not the only concern.

By the middle of the 1980s, food scares steered consumers toward questions of not just food processing, but also conventional farming practices themselves, which increasingly became seen as dangerous. One of the first scares involved the fungal fumigant ethylene dibro-mide, or EDB. In 1983, the Environmental Protection Agency (EPA) began an investigation after the state of Florida banned the sale of twenty-six products that contained high levels of the cancer-causing chemical. As consumers worried about the safety of the products in question, "Some of the companies whose food products were ordered removed from the market in Florida [began] contemplating legal ac-tion against the state" (Shabecoff 1983). This carcinogenic pesticide was also found on citrus products, leaving the Florida citrus growers worried about possible bans, while citizens learned their wells also had been contaminated (Rosenbaum 1984). Stories such as these sent a clear message to many skeptical consumers: the conventional food industry had become unable to ensure the safety of its products.

A mere three months after Florida banned the sale of products with high levels of EDB, shoppers across the country had grown genuinely concerned about food safety. According to a survey by the Food Marketing Institute in March 1984, "77 percent of those polled considered residues such as pesticides and herbicides a serious hazard, while cholesterol was considered a serious hazard by 45 percent, followed by salt, 37 percent" (Associated Press 1984). It is interesting to note that the survey was conducted before the EDB controversy began. Pesticides were perceived as a greater health risk than the now usual suspects of cholesterol and salt. Commonplace agricultural chemicals were becoming a national issue.

However, pesticides were not the only concern. In September 1984, "For the first time," doctors reported to have traced "a serious outbreak of human food poisoning to drug-resistant germs that spread from beef cattle routinely fed antibiotics to promote growth" (Haney 1984). Doctors were able to trace the strain of antibiotic resistant salmonella that infected eighteen people in the Midwest to a herd of cattle from South Dakota. These cases stand in stark contrast to traditional food scares. They were not the result of an accidental contamination or an equipment malfunction—such as, for example, when canned goods are not properly heated, leading to botulism. Far from it; these cases involved the "best practices" of industrial agriculture—practices designed to produce food cheaply and efficiently, despite the human cost.

Throughout the 1980s, food scares became a part of everyday life in America. Salmonella, trichinosis, E. coli, botulism, and pesticides all gained the status of common parlance. One of the most dramatic scares came at the end of the decade. In February 1989, the EPA announced that the chemical daminozide, sold under the brand name Alar, was a carcinogen and began taking steps to remove it from the market. Alar was used primarily on apples to enhance the color and uniformity of the fruit. Although the chemical was used on as few as 5 percent of the nation's red apples (Leary 1989), the public response was unprecedented.

A report by the Natural Resource Defense Council revealed that "the average preschooler's exposure to this carcinogen is estimated to result in a cancer risk 240 times greater than the cancer risk considered acceptable by the EPA following a full lifetime of exposure" (Oakes 1989). Consumer panic ensued. "School systems in Atlanta, San Francisco, Chicago and dozens of other cities stopped distributing apples," following the examples set by New York City and Los Angeles (Associated Press 1989a). To support prices, the USDA bought $15 million in surplus from the apple industry with the intention of providing the apples to school lunch programs, prisons, and food aid programs (Associated Press 1989c). Many farms that used the chemical were foreclosed, and apple processors saw sales plummet. Almost overnight, conventionally grown apples went from being a healthy snack to a toxin.

As the safety of America's food supply came into question, organic food stood out as the perfect alternative. Produced with ecological production processes instead of chemical inputs, it did not trigger concerns about chemical residues. As a result, organic food entered the mainstream of American dietary consciousness for the first time. As unsafe food was discovered in nearly every corner of the grocery store, consumers resisted its advance. Organic food, once the stuff of iconoclasts, anti-capitalists, and hippies, became part of the arsenal of concerned parents. Organic food became the safe alternative food.

The growing appeal of the organic food movement was becoming apparent. In 1989, a national survey revealed that 84 percent of respondents would buy organic if given the choice, and nearly 50 percent said they would be willing to pay more (Associated Press 1989b). By 1990, U.S. organic food sales were estimated to have reached $1 billion (Organic Trade Association 2011). However, organic food was still an alternative. The 1980s represented a new wave of resistance, a broadly based consumer resistance to the perception of conventional foods as unsafe. And corporations still remained hostile to challenges. As the case of EDB revealed, agribusinesses preferred to sue rather than admit its practices were unsafe. The Alar scare provoked a similarly critical response from the chemical industry (Oakes 1989), and the chemical was abandoned only because consumer resistance was so immense. Organic food was still a challenge to the status quo, but not for long.

From its inception in the 1920s to its flourishing in the 1960s, organic farming was an attack on the place of industry in the life of the farm and on the table. To be certain, the organic movement was part of a dynamic historical change. The earliest organic pioneers were social conservatives, resisting the corrupting effects of artificial fertilizers—not only on the soil, but on society as well. During the countercultural movement, the organic movement was logically incorporated into the environmental movement, as the effects of artificial agricultural inputs on the environment became all too apparent. Finally, over the course of the 1980s, the organic movement expanded as everyday Americans attempted to resist the seemingly unending list of unsafe products making their way onto the dinner table. Despite these transformations, the organic movement remained fundamentally oppositional.

Such was the organic movement, but what has it become? How did a movement that defined itself as alternative become part of the conventional food system, just another component of agribusiness? How did "organic" become "mainstream"?

In most accounts of the organic movement, 1990 stands out as a landmark year. It is the moment in which the entire history of the movement pivots, alternately described as the year of the organic movement's rise or fall, depending on whom you ask. By 1990, the organic movement was roughly seventy years in the making, with roots stretching back to the 1920s. How can a history of resistance spanning decades rise or fall on the basis of the events of a single year?

Of course, accounts that posit 1990 to be a watershed moment in the organic movement are incomplete. We should remember that the events of 1990 are the culmination of decades of activism and resistance, as well as decades of agro-industrial growth and corporate attempts at co-optation. Nevertheless, 1990 is a critical point in our story of the organic movement. Simultaneously seen as both a blessing and a curse, it is the definitive point at which the organic food movement hit the mainstream. With the passage of the U.S. Organic Foods Production Act (OFPA) of 1990, organic food left the fringes of America's agricultural economy and received federal recognition—and regulation.

Along the way, organic movement NGOs and activists were at the table, fighting to gain recognition for organic farming principles and practices through federal standards. Nevertheless, many scholars have highlighted the relative power of other actors at the table—namely, agribusiness interests—to co-opt the meaning and practices of organic farming (DuPuis and Gillon 2008; Guthman 2003; Jaffee and Howard 2010; Johnston, Biro, and MacKendrick 2009). This is not to suggest that the federal organic standards were a corporate takeover plot or that they were foisted onto organic movement actors. Indeed, throughout the regulatory process, federal standards were endorsed and accepted by members of the organic farming community. Nevertheless, scholars have demonstrated the ways the final implementation of federal organic standards often silenced movement voices (DuPuis and Gillon 2008) and fundamentally transformed the structure of American organic farming by fostering both the entrance and ascendance of industrial actors into the organic sector (DeLind 2000; Guthman 2004c). It is these changes—and the actors who possessed disproportionate power to effect change in the organic market—that are our primary concern here. It is these changes that helped create a structurally bifurcated organic market, and as a result, it is these changes that provide the context in which today's organic farmers are working to achieve good matches.

Despite the outcome, when viewed in a vacuum the OFPA seems relatively benign. The congressionally mandated purpose of the act is quite straightforward: to create national standards governing organic products, assuring consumers that organically produced foods meet a consistent standard and facilitating interstate commerce in fresh and processed foods that are organically produced (OFPA 1990, Title XXI, §2102). However, the development of the act can also be construed as the result of various powerful social forces operating within and upon the organic movement.

What is it about the regulation of the OFPA that so strongly signifies the incorporation of organic agriculture into the "mainstream" food production system? When you think about it, most aspects of our lives are regulated in some way. The roads we drive on, the medicines we take, and the restaurants at which we eat are all regulated. There is seemingly no way to escape government oversight in our lives. Were we to lock ourselves at home to avoid interacting with government regulations, we would be doomed to failure—even our homes are circumscribed by habitability regulations. As a result, the regulation of organic foods is hardly unique. To understand its profound significance, we will have to examine the processes and pressures that brought us to this watershed moment and where we have come since then.

Regulating Organic Farming

The critical significance of 1990 in the history of organic agriculture is that in that year the federal government took a definitive stand as to what constituted "organic farming." The OFPA was not regulating the organic *movement*. Rather, it set out to institute a narrow vision of organic *agriculture* based on "consistent standards" to facilitate "interstate commerce" (OFPA 1990, Title XXI, §2102). Such standards were deemed necessary because, up until this point, organic agriculture existed in what can be called an "unsettled" market—"unsettled" because categories, values, and commensurability were not clearly demarcated. As we have seen, the organic movement ran the gamut of philosophical positions from conservative to liberal, Luddite to eco-apocalyptic. Such philosophical pendulum swings are hardly a stable foundation for the commensurability necessary for economic transactions.

Contrary to typical reductionist accounts, the OFPA was not the federal government's first examination of organic farming policies. Rather, government involvement in the organic movement began in the early 1980s, following the wave of consumer interest in organic products generated by the counterculture, as well as by food scares. The first comprehensive examination of organic farming came ten years before the presumed watershed moment in 1990, when the USDA issued its "Report and Recommendations on Organic Farming" in 1980.

The foreword to the report, written by Secretary of Agriculture Bob Bergland, reveals the social forces behind the federal government's interest in organic agriculture. Secretary Bergland writes, "We in USDA are receiving increasing numbers of requests for information and advice on organic farming practices. Energy shortages, food safety, and environmental concerns have all contributed to the demand for more comprehensive information on organic farming technology" (cited in USDA 1980, iii). In other words, many of the forces driving the increasing legitimacy of organic agriculture were linked to social issues external to the movement. Nevertheless, such forces were increasingly propelling the movement into the national political—and economic—spotlight.

The 1980 report relied on a holistic definition of organic farming, describing processes rather than inputs. For the purposes of the report, organic agriculture was a system where "to the maximum extent feasible, organic farming systems rely upon crop rotations, crop residues, animal manures, legumes, green manures, off-farm organic wastes, mechanical cultivation, mineral-bearing rocks, and aspects of biological pest control to maintain soil-productivity and tilth, to supply plant nutrients, and to control insects, weeds, and other pests" (USDA 1980, xii). Harkening back to the earliest organic pioneers, this definition outlined practices of nourishing and replenishing the entire system, not just the crops. Further, the report recognized that the organic movement was held together by an overarching philosophy allowing for a diverse array of interpretations.

As a result of this process-based production system, organic farming was described as representing a "spectrum of practices, attitudes, and philosophies" (USDA 1980, xii). In fact, the 1980 report did not even begin with an *a priori* definition of organic agriculture. Instead, the study authors sought to "reflect the essential agronomic components and characteristics of organic farming technology discovered over the course of the survey" of sixty-nine participating farms (Youngberg and DeMuth 2013, 303). As a result, farms that would not be considered certified organic today, due to their use of limited chemical inputs, were included in the report's definition.

At the time of the USDA's 1980 report, organic farming was hard to pin down. In fact, it always had been; in 1946, Sir Albert Howard declined to join the Soil Association out of concerns that it was not prohibitive enough in its approach to chemical usage (see Reed 2001). Moreover, the 1980 report's goals were far more sweeping than the creation of a singular definition of organic farming. Instead, the USDA commissioned the report "to learn as much as possible about how organic practices might be incorporated into conventional agricultural production systems as a means to help alleviate some of the problems that had begun to plague America's agriculture" (Youngberg and DeMuth 2013, 303). Indeed, there was great diversity under the canopy of "organic." Nevertheless, the guiding principles of the movement, such as sustainability and the stewardship of farms as ecological systems, were recognized as giving shape to these disparate practices.

While the commissioning of the 1980 USDA report reflected the increasing legitimacy of organic farming as an alternative system of agriculture, it did little to settle the market definition of "organic." Regardless, this increased legitimacy was short-lived, at least within the federal government. Secretary Bergland, a farmer from Minnesota whose approach as secretary of agriculture was influenced by the success of his neighbor's 1,500-acre organic farm, was replaced after the election of President Reagan in 1981 by John Block, a large-scale hog and grain farmer who opposed the report in its entirety (Youngberg and DeMuth 2013). Under the leadership of Secretary Block, "the Report was disowned by the USDA's new leadership and the term 'organic farming' officially became taboo (again)" (Lipson 1997, 20). While many had hoped the report would mark a shift in the USDA approach to organic agriculture, it only inflamed its critics with its focus on an " 'ideology' of environmental responsibility" and calls for " 'holistic' research and education" (ibid.).

Others within the USDA who had been involved with the 1980 report would also be replaced under the Reagan administration. Garth Youngberg, the leader of the USDA Organic Study Team, which produced the 1980 report, went on to become the USDA organic farming coordinator, a position created as a result of the report's findings. Merely a year later, and still under the Reagan administration, he was directed to spend only half of his time following up on activities related to the 1980 report. A year later, Youngberg was fired, and the position done away with (Youngberg and DeMuth 2013).

Fearing the Emergence of Organic: From Agricultural Reform to Consumer Choice

Just two years after the "Report and Recommendations" was produced, the Organic Farming Act of 1982 was proposed in the U.S.

House of Representatives to establish six regional research centers to improve farmers' access to knowledge about organic agriculture. The USDA, however, decided not to support the bill, claiming that such research was beyond its budget (League of Conservation Voters 1983, 19–20). During the congressional hearings before the Subcommittee on Forests, Family Farms, and Energy, Dr. Terry Kinney of the USDA's Agricultural Research Service testified that the department was "sympathetic" to the desires of the proposed bill. However, during questioning, several senators expressed doubts. For example, Representative George Brown of California suggested that he was "disturbed at the fact that there is some indication that a counter reaction against organic farming may be setting in" after the cautious but positive appraisal of organic practices in the 1980 "Report and Recommendations on Organic Farming" (U.S. Congress 1982, 14). Brown proceeded to question Dr. Kinney about possible attempts to quash the 1980 report and earlier allegations that the USDA had suppressed favorable studies of organic farming.

Such tensions surrounding organic agriculture within the USDA intensified throughout the hearing. For example, Representative Tom Daschle of South Dakota referenced a report in *Ag Consultant* magazine that quoted Agriculture Secretary Block assuring that "there would be no follow-up by the present administration on this dead end type research"—"dead end" referring to organic farming research (U.S. Congress 1982, 14). While Dr. Kinney asserted that such an assurance was incompatible with the USDA's official position of being "sympathetic" to the proposed bill, Daschle pressed that it was perfectly compatible with the testimony Kinney had provided. The USDA was unwilling to allocate its budget to a method of farming it had reported favorably upon only two years prior. Representative Daschle concluded, "What you are telling us in your statement is one thing; what you are telling us in answer to these questions is something totally different" (16).

With the political changes brought by the Reagan administration, the federal approach to organic agriculture changed dramatically. The USDA's 1980 report had sought to expand research and support for organic farming principles so that they could be integrated into the agricultural mainstream, in an effort to reduce the use of environmentally harmful practices across the U.S. agricultural sector. Under the Reagan administration, the 1980 report was rejected, and debates about organic farming shifted to the types of definitional concerns necessary to settling a niche market. Such a shift was entirely consistent with the broader agricultural projects of neoliberalism undertaken in that era.

Under the Reagan administration, the goal for agriculture in general was to "unleash market mechanisms in agriculture by limiting government's influence over production decisions and market prices" (Winders 2009, 160), not to create a new set of standards for the entire U.S. agricultural system, as the 1980 report had suggested was necessary (Youngberg and DeMuth 2013, 303). Disavowing the approach to organic outlined in the 1980 report, the Reagan administration advanced an agenda favoring policies fostering market solutions and "consumer sovereignty" through increased choice. Reframed as a niche market, organic would come to represent an alternative for those willing and able to afford market premiums, leaving consumers generally with a less regulated conventional food system (Guthman 2011). While the Reagan administration woefully failed in reducing government spending in agriculture (reaching, in fact, a national record of nearly $26 billion in 1986), it did succeed in aiding the creation of a lucrative new market for agro-industrial firms by pushing for an industry-friendly definition of "organic."

Within the 1982 hearings, even organic movement NGOs suggested there was an incipient need for a narrowly defined, input-based definition of organic farming. For example, the coordinator for the Carolina Farm Stewardship Association, Debbie Wechsler, testified, "Perhaps the only distinction that can possibly be made is that 'organic farming' does not use manufactured chemicals such as fertilizers and pesticides. 'Conventional farming' has used and continues to use crop rotation, legume nitrogen, animal wastes and agricultural limestone in addition to other available technologies. In the continuum of available technologies, where is the line drawn to create an ideology separate from other crop production practices?" (cited in U.S. Congress 1982, 118).

Wechsler's testimony problematizes nearly every characteristic of organic farming listed in the USDA's 1980 report. A process-based model of understanding organic farming—precisely the model developed over the course of the movement—could not create a settled market for the agricultural industry. There was simply too much ideology and too many diverse practices in the way. As Wechsler testified, "The choice of a rotation containing the forage legume alfalfa is independent of ideology" (cited in U.S. Congress 1982, 119). In order to gain government support—particularly in the form of research and programs for alternative farming through the Cooperative Extension System—many in the organic movement recognized a need to settle on a concrete definition of organic methods.

As a result, the debates and regulatory failures of the 1980s laid the foundation for the development of the landmark 1990 OFPA. Over the course of the 1980s, organic agriculture continued expanding for a number of reasons. First, the countercultural movement of the previous decade had revitalized the older organic movement and infused it with contemporary cultural significance. Further, economic conditions such as energy shortages and rising costs of chemical inputs were squeezing the profit margins of conventional small farms more tightly. Simultaneously, the market for organic foods was exploding in light of consumer concerns about the environment and fears about the health and safety of conventional foods.

Amid such rapid growth, the organic market still remained fundamentally unsettled because the process-based definition of the movement did little to produce commensurable products that could be exchanged in the larger agricultural economy, and the patchwork of local and regional standards crafted by organic movement actors left room for inconsistencies. As a result, the landmark 1990 act can be seen as an attempt by the federal government to settle the market by introducing consistent, rational standards that could replace nebulous processes based on philosophical assumptions. As DeLind has noted, "Agribusiness leaders, environmentalists, consumers, processors, and organic farmers themselves" worked for and supported federal standards, believing that "multiple standards permit too much variability within the industry and too much consumer confusion" (2000, 199). To put it another way, federal standards were a way of making an alternative system of production and consumption a niche market within the agricultural mainstream.

Rather than embracing a view of organic farming as a "spectrum of practices, attitudes, and philosophies" (USDA 1980, xii), the crafting and implementation of the 1990 OFPA took a decidedly input-based approach to defining organic foods. And despite the efforts of many NGOs and organic farmers to gain federal standards, some of these individuals began to feel disenfranchised. For example, DeLind quotes an organic farmer who explained, "Organic farmers felt they won a victory when the 'new' farm bill made a provision for a national organic certification program, but that is when I sadly said goodbye to the Organic Movement" (2000,199). To be considered organic, the fundamental requirement of the OFPA is that a food be "produced and handled without the use of synthetic chemicals" (OFPA 1990, Title XXI, §2105). In this way, some argued that the nuance of the organic movement was reduced to the simple avoidance of synthetic chemicals—all of the philosophy, countercultural values, lifestyle choices, and concerns for community became incidental (DeLind 2000; Guthman 2004c). Of course, avoiding chemicals can be accomplished by means of philosophically guided processes but it need not be, and that is precisely the point. While diversity of processes made economic commensurability tenuous, the argument goes, the focus on the exclusion of chemical inputs settled the market at the organic movement's "lowest common denominator" (Howard 2009, 14).

In the entirety of the 1990 OFPA, the notion of organic farming involving a spectrum of process-based practices appears only once, in a relatively small section on the "Organic Plan." According to the act, every farm must have a plan detailing maintenance of soil fertility, manure usage guidelines, livestock considerations, handling measures to ensure organic integrity, and management of wild crops

(OFPA 1990, Title XXI, §2113). However, there is no description of a particular framework to guide the planner. As we have seen in the 1982 debates, these processes are not, in and of themselves, inconsistent with conventional farming practices. It is the ideology behind the practices—the holistic agro-ecological approach to agriculture—that historically distinguished organic farming from its conventional counterpart. All an "organic plan" must do, in addressing the requisite components, is to not undertake practices inconsistent with the act—in other words, no chemicals, please (§2102).

Once again, in order to mainstream organic farming, it came down to inputs. As a result, the OFPA established a process to determine which substances could be used in organic production and which could not. In the interest of consistent standards, the USDA was charged with the creation of the National List of Allowed and Prohibited Substances. The purpose of this list was to eliminate the guesswork involved in determining what was really organic by specifying approved and prohibited substances (OFPA 1990, Title XXI, §2118). With the development of the list, a rationalized litmus test for inclusion within the organic farming community was established. If you avoided the prohibited chemicals, you were in the clear.

Again, this standard was not intrinsically at odds with the goals of organic farmers and the broader organic movement. These groups had actively advocated for federal organic standards and by the 1990s had already worked to craft forty state and regional certification programs (Jaffee and Howard 2010). Indeed, "Organic growers have long recognized the need for standards to ensure the integrity of their methods and products" (DeLind 2000, 198). However, as the 1990 OFPA was gradually implemented, many within the organic movement resisted what they felt had become a race to the bottom.

A Process of Avoidance in Mainstream Organic

The drafting of the Organic Foods Production Act of 1990 was an effort to proactively regulate a system of farming in the process of transitioning from an alternative lifestyle to an emergent industry.

Moreover, the act was not the last word on defining what qualified as "organic." In the course of developing the final regulations, the language of "process" worked itself back into the discourse. For example, in 1995 the National Organic Standards Board defined organic agriculture as "an ecological production management system that promotes and enhances biodiversity, biological cycles and soil biological activity" (Gold 2007). However, as regulations were drafted, it soon became apparent that an input-based approach to organic agriculture was still operating behind the veneer of revised language.

Despite this input-based approach, the process of developing federal organic standards should not be seen as merely a story of corporate influence diluting the meaning of the term "organic." In the eyes of many activists and scholars, such was the result. Joan Gussow—who served as an adviser and consumer/public interest advocate on the National Organic Standards Board from 1996 to 2001—suggested, "While sustainable agriculture cannot be defined, organic agriculture *is* being defined, its definition being rendered serviceable to an existing agrifood industry" (cited in DeLind 2000, 199). Nevertheless, the process was contentious and one to which activists, farmers, and consumers contributed and—to some extent—even redirected. Nowhere is this clearer than in the case of consumer resistance to the USDA's proposed standards issued on December 16, 1997.

In the USDA's proposed standards, organic producers were permitted to use "food irradiation for controlling spoilage and bacterial contamination, sewage sludge as a soil amendment, and genetically modified organisms," along with a host of other departures from the principles many within the organic movement hoped would be safeguarded by federal oversight and regulation of the expanding organic sector (DeLind 2000, 199). The response of consumers, farmers, and movement NGOs was unprecedented. During the public notice and comment period, over 275,000 comments were filed. Under this intense criticism, the USDA withdrew the proposed standards.

In the wake of the public resistance to how organic agriculture was being defined, former deputy secretary of agriculture Kathleen Merrigan played a critical role. Merrigan was both responsible for drafting the OFPA under Senator Patrick Leahy and also oversaw the revision and implementation of the OFPA standards as administrator of the USDA Agricultural Marketing Service. As a result, Merrigan "is commonly referred to as the 'midwife of the National Organic Program' for her proactive role in managing the process" (Lohr 2009, 2). Even after the implementation of national organic standards, movement activists were hopeful that Merrigan's 2009 appointment as deputy secretary of agriculture by President Barack Obama might crack what has been seen as the USDA's single-minded approach to organics (ibid.), a testament to the lasting influence of her role in revising the organic standards and eliminating the most grievous departures from organic farming ideology and practice.

It was not until 2000 that the USDA issued the organic standards mandated in the 1990 statute, in a final form made more palatable thanks in part to the types of revisions Merrigan worked to oversee. Nevertheless, in the final regulations adopted by the USDA, when it comes to products being federally recognized as organic, a lot still comes down to the National List of Allowed and Prohibited Substances. Mandated in the 1990 OFPA, the National List was intended to serve as the boundary for the new, mainstream organic agriculture. However, by the time the final rules were implemented in 2000, numerous synthetic substances had already been exempted. Many of these exemptions were intended to either ease issues of fertilization and pest management or to allow the manufacture of organic processed foods, which required ingredients that were either unavailable in sufficient quantities in organic form or were critical for preserving organic processed foods for national distribution.

Some of the exemptions for farming operations seem relatively benign and practical—for example, an exemption for newspapers (without colored or glossy inks) for use as mulch (OFPA 1990, Title VII, §205.601). Others, however, allow for the use of more controversial compounds. For example, the National List provides an exemption for copper sulfate (§205.601), which can be used to control bacterial and fungal diseases. Although the EPA classifies copper sulfate as a Class I pesticide, indicating it is highly toxic (Extension Toxicology

Network 1996), the National Organic Program finds it acceptable as long as it is "used in a manner that minimizes accumulation of copper in the soil" (OFPA 1990, Title VII, §205.601).

While the toxicity of copper sulfate would seem to suggest that its accumulation in the soil would be a serious concern, the National List leaves its use seemingly open to each farm operator's interpretation. The fact that copper sulfate is exempted in the National List—and, therefore, is technically organic—doesn't mean it is any safer than other chemical fungicides. Copper sprays have been shown to "endanger" honeybees, and copper sulfate runoff is "highly toxic" to fish and aquatic invertebrates (Caldwell et al. 2005, 91–92). Reflecting on the serious dangers of copper accumulation in the soil, a toxicology report noted that "most animal life in soil, including large earthworms, has been eliminated by the extensive use of copper-containing fungicides in orchards" (Extension Toxicology Network 1996).

Along with the various chemicals allowed in the fields, other exemptions granted for organic food processing further complicate the current meaning of organic. Non-organic hops, for example, are officially permitted in the brewing of organic beer (OFPA 1990, Title VII, §205.606). This exemption is permitted since hops comprise less than 5 percent of a bottle of beer, making them a minor ingredient in the product. The case of hops, however, has received sustained attention because "most other minor ingredients on the National List . . . tend to be food-based non-organic flavorings and food colors (such as carrot-based annatto), rather than a central ingredient in the production of the substance" (DuPuis and Gillon 2008, 51). Other notable exceptions include intestinal casings for sausage, Turkish bay leaves, chipotle chilies, frozen lemongrass, and domestic cornstarch (§205.606). Apart from such exempted agricultural products, forty-four synthetic chemicals are officially exempted from the National List and are allowable in products labeled as organic (§205.605b).

For many supporters of the organic movement, the accommodations to conventional agribusiness methods were all too apparent. Not only had the movement's message been narrowed down from a system of farming to a list of excluded chemicals, but many chemicals were also making it into mainstream organic products. More-over, if ingredients like chipotle chilies, sausage casings, and domestic cornstarch—all of which could be organically produced—were exempted for organic food manufacturers, what did the organic label even mean? If we consider the forty-four exempted synthetic chemicals allowed in federally recognized organic products, formulating an answer to that question becomes even more tenuous.

True, the exceptions are for "minor ingredients," and the exempted agricultural products are permissible only when the product is not "commercially available in organic form" (OFPA 1990, Title VII, §205.606). However, the logic seems rather circular and self-serving. There is no reason why frozen lemongrass, chilies, or domestically produced cornstarch should not be commercially available in organic form. If organic products were required to include only organic agricultural components, wouldn't that create a market for them? By the creation of such exemptions, many argue, the value of organic alternatives is weakened, in a sense ensuring they will not become commercially viable—and therefore available (DuPuis and Gillon 2008). In the case of hops, "It is no longer necessary for any brewer to buy organic hops for the production of organic beer, even though smaller producers had previously been using organic hops as an input in their products" (51). The legislative "mainstreaming" of organic did more than transition organic farming from a marginal set of alternative practices to a federally recognized niche market within the agricultural mainstream; it actively sought to incorporate aspects of mainstream farming practices—and even mainstream agricultural products—into the purview of the organic label.

In 2002 Arthur Harvey, an organic blueberry farmer and licensed certifier from Maine, filed a lawsuit as a consumer against the USDA, alleging that the organic standards the agency adopted were inconsistent with the clear intent of the 1990 OFPA (DuPuis and Gillon 2008, 47). If the 1990 legislation was clear about anything, it was that synthetic substances could not be included under the organic label. Although initially defeated, on appeal Harvey prevailed on three counts. As the *New York Times* reported, the decision "sent shivers down the spine of many organic food manufacturers" (Warner 2005). If anything, the decision revealed how contentious the process of establishing a new market category could be. Transforming ideological categories into economically viable products is by no means an automatic process; it takes work.

Recently, the Harvey case has received scholarly attention. DuPuis and Gillon have argued that the Harvey case "shows us a civic struggle between parties defining organic as a standard and those who had a more deliberative definition of organic as an alternative mode of governance" (48). They stress the "deliberative process" of "boundary work" needed to maintain organic food as a commodity that consumers are willing to buy at a premium. Not only was the historic definition of organic farming based on a spectrum of practices, the creation of the meaning of organic was a process of contestation, of resistance.

The Harvey case was essentially an attempt to challenge the exclusive legislative definition of organic. It is an example of boundary work, of a collective process of contesting what is and what is not included in the category of organic food. Despite Harvey's victory, it was only one battle. Nor was it a sign that organic boundaries would remain open to the boundary work of the organic movement. Guided by the Organic Trade Association, an organic industry lobbying organization, Congress included legislation in the 2006 budget for the USDA that clarified the "inconsistencies" in the 1990 OFPA that were the foundation of Arthur Harvey's legal case. In resolving these inconsistencies, Congress "allowed synthetics back into organic food processing by overturning Harvey's third count and authorized 'emergency procedures' for adding non-organic agricultural products into organic foods" (DuPuis and Gillon 2008, 50; see also Jaffee and Howard 2010). This monumental addition to a federal budget, classified by the Organic Trade Association as "a discreet, very limited, legislative action," became known as the OTA Rider (DuPuis and Gillon 2008, 49).

The OTA Rider is critically important to our story for several reasons. First, it reveals the radical transformation of organic agriculture that occurred as the philosophical heart of the movement was stabilized into a commensurable market category. Historically, organic agriculture was an effort by farmers to opt out of the larger food system. It was a way to resist industrialization, environmental degradation, and unsafe farming practices (Jaffee and Howard 2010). While organic advocates since Steiner and Howard have eschewed chemical solutions to agricultural problems, organic farming signified more than merely avoiding certain chemicals. The organic pioneers were not willing to unquestioningly accept the growth of modern agriculture. Organic farming was intended to represent an alternative vision of agricultural growth, concerned with growing better—not merely more.

However, as the Organic Trade Association lobbied Congress to establish the boundaries of organic farming, it argued that "market led growth is only possible if organic farmers and processors compete on level ground with non-organic farmers and processors" (OTA 2005). In the OTA Rider, as well as the earlier OFPA, organic farming was recast as a market niche in which the agribusiness mainstream could participate—a redefinition fundamentally at odds with "the damning ecological critique of industrial agriculture that guided the organic pioneers" (Jaffee and Howard 2010, 397). As Walmart executive Bruce Peterson would later suggest, "Organic agriculture is just another method of agriculture—not better, not worse" (cited in Warner 2006). With the regulation of organic agriculture, organic farming was recast simply as a way for corporations to provide consumers what they wanted—a unique product, but not a better product.

Further, the OTA Rider reveals the increasing exclusion of members of the organic farming movement in determining the "real" meaning of their practices. Iowa Democrat Tom Harkin made the following remark on the Senate floor, chastising his colleagues for the inclusion of the OTA Rider in the budget appropriations bill through closed-door committee meetings: "Behind closed doors and without a single debate, the Organic Foods Production Act was amended at the behest of large food processors without the benefit of the organic community reaching a compromise. To rush provisions into the law that have not been properly vetted, that fail to close loopholes, and that do not reflect a consensus only undermines the integrity of the National Organic Program" (cited in DuPuis and Gillon 2008, 49). Harvey, the Maine blueberry farmer who sued to keep the USDA accountable, was not included. Congress negated the judicial decision for which Harvey had fought.

If the organic movement was no longer setting the boundaries of what qualified as organic, it was because larger business interests had pushed it out. According to a report in the New York Times, "Dean Foods' subsidiary Horizon Organic and the J. M. Smucker Company, the owner of Knudsen and Santa Cruz Organic juices, said they supported the work by the Organic Trade Association" (Warner 2005). Kraft's majority owner, Altria, even had lobbyist Abigail Blunt—wife of the interim House majority leader at the time of the OTA Rider's passage—working on the issue (ibid.). Who better than agribusiness representatives to usher organic farming into the mainstream agricultural fold?

The history of organic farming is one of ever-changing challenges to America's agribusiness. However, stabilizing the definition of organic farming into a generic input-avoidance regime threatened the legitimacy of the original concept of organic. The transformation of organic into a market category was not an automatic process—far from it. The foundations were constructed in the debates of the 1980s, which laid the groundwork for what would come. The landmark Organic Foods Production Act was hardly a watershed moment. It was born out of decades of organic activism, growth, and ultimately incorporation of organic into the larger system of farming. Nor was 1990 the definitive moment it is often made out to be. Keeping the market category of organic food "settled" would continue to take work in the years to come, leading to the passage of the OTA Rider and the increasing recasting of "organic" as a niche category of America's larger food system.

Nevertheless, by establishing a commensurable definition of organic food products based on inputs, the federal government paved the way for the expansion of the organic market. Following the government's official recognition of organic agriculture, the organic industry entered into a period of tremendous growth across all measures. Beginning in 1990, organic retail sales grew by at least 20 percent annually. After 2000, steady growth continued, only slumping in the wake of the 2008 economic recession. The land devoted to organic cultivation also rapidly increased after 1990. In fact, the acreage devoted to organic farming doubled between 1992 and 1997, reaching 1.3 million acres. The growth of some sectors of the organic market was even more astounding, with organic dairy sales rising by upwards of 500 percent between 1994 and 1999 (Dimitri and Greene 2002, iii).

Where has all of this contestation led us? Some would argue it has led us to a new era of organic agriculture: Supermarket Organic, available at a Walmart, Whole Foods, or nearly any other supermarket near you.

According to a USDA report, the organic industry experienced a landmark moment in 2000. "For the first time," its authors stated, "more organic food was purchased in conventional supermarkets than in any other venue" (Dimitri and Greene, 2002, iii). "Supermarket organic" represents not only organic foods produced in farming systems with the level of intensity and scale necessary to get them onto the shelves and into the produce aisles of chain grocery retailers, but also the entire set of social relations such a food system fosters. It is the new norm in the United States, and this change is profoundly important. The social, economic, and environmental relations fostered by the organic products the majority of Americans are now accustomed to seeing on grocery store shelves represent a significant shift from both the priorities of the historic organic movement and the beliefs and practices of many contemporary small-scale farmers committed to an alternative vision of agriculture. As organic agriculture came to be defined primarily in terms of prohibited chemical inputs rather than as a broader, agro-ecological and community-based process, the terrain of the organic market shifted in ways that benefited the types of large organic operations that could meet the demands of an industrial food system. For some, the transformation from a small movement into a formidable industry has led to a "process of establishing and enforcing standards [that] has de-emphasized or ruled out of bounds" more sustainable visions of organic (Jaffee and Howard 2010, 397).

The history of the organic movement up to the present day reveals that the ideologies, ethics, and practices of organic activists have shifted over time. In the nearly two centuries of modern organic farming, there is no historical consensus on many of the issues surrounding organic food production. However, the current regulatory framework is unprecedented in its exclusive focus on chemical inputs. This narrow focus ignores broader concerns about the environmental, economic, and social sustainability of our food system, concerns that in one form or another have motivated alternative visions for organic farming.

Before we consider the shifting terrain of the contemporary organic sector, it will be useful to once again reflect on what could constitute a more expansive notion of alternative food systems than the narrow, institutionalized definition of organic that today's farmers and consumers have inherited. Most agro-ecologists agree that a truly sustainable alternative agriculture cannot simply be defined by the absence of synthetic chemicals (Altieri 1995). Rather, an alternative to the current paradigm of industrial food production must at least incorporate broad concerns about environmental, economic, and social sustainability (Agyeman and Evans 2004; Alkon 2008; Feenstra 2002; Follet 2009; Gillespie et al. 2007; Gliessman 2006). Given the multitude of situations a sustainable food system must address, it may be neither possible nor beneficial to prescriptively define the practices of sustainable agriculture (Hinrichs 2010). However, how have the various aspects of agricultural sustainability been conceptualized?

Consonant with views about the nature of organic farming before it was defined through regulation, such as those found in the USDA 1980 "Report and Recommendations on Organic Farming," environmentally sustainable agricultural systems require an agro-ecological approach rather than a well-defined set of practices (Pretty 1998). This means that environmentally sustainable practices in agriculture have been cast as a matter of "replacing external inputs with more site-specific agro-ecological processes that reduce environmental harm and depletion" (Hinrichs 2010). John Ikerd has articulated a definition of organic agriculture that captures the notion of environmental sustainability well: "Organic farming methods are based on nature's principles of production—on farming in harmony with nature rather than trying to conquer nature. Diverse organic farming systems, most of which integrate crops and livestock enterprises, are designed to capture solar energy, to recycle waste, and to regenerate the health and fertility of the soil. Organic farmers see themselves as stewards of nature" (2001). Such an approach views agricultural operations as holistic ecological systems rather than a reduction of farming to a matter of managing inputs. Unfortunately, while the organic movement has focused much of its attention on what it means for farms and farmers to promote environmental well-being (Buttel 2006), this focus has often come at the expense of a more holistic focus on social and economic aspects of agricultural sustainability (P. Allen 2004).

In order for the contemporary food system to be truly sustainable, however, social and economic concerns must be of paramount importance (P. Allen 2004; Johnston, Biro, and MacKendrick 2009). Indeed, a truly sustainable food system might require trade-offs when the most environmentally sustainable option is in conflict with other sustainability goals. "Does an environmental criterion for sustainable food systems trump the importance of more direct relationships between farmers and consumers? If it does, [a consumer] may need to reconsider meandering [through] the countryside in [a] private automobile to an array of local markets and farms where [he or she can] experience that farmer connection. But who decides? Whose values matter?" (Hinrichs 2010, 25). Even if strict guidelines for attaining agricultural sustainability are untenable given the shifting and emergent conditions a sustainable food system must address, many principles for social and economic sustainability have been identified (Hinrichs 2010; Kirschenmann 2004). These include concerns for the social and economic viability of local communities (Gillespie et al. 2007; Lyson and Green 1999; Schor 2010); concern for the rights and welfare of farm workers (P. Allen 2004; Harrison 2011); the redress of the raced, classed, and gendered inequalities produced by the contemporary food system (P. Allen 2004; Alkon 2008; Alkon and McCullen 2011; Block et al. 2011; Guthman 2008a, 2011; Johnston and Baumann 2010; Johnston and Szabo 2011; Trauger et al. 2010); improvements in social connections (Feegan and Morris 2009; Gillespie et al. 2007; Schor 2010); and the cultivation of opportunities for "citizen-based" modes of engagement in the food system (Chiffoleau 2009; Johnston, Biro, and MacKendrick 2009; Seyfang 2006).

Given these criteria for a more sustainable agricultural economy, it will be useful to first examine where the history of the organic movement has taken us before moving on to interrogate whether the mainstream organic movement we have inherited is capable of delivering a sustainable organic future. To be sure, the redefinition of organic food in economically rational terms that occurred amid the regulatory debates of the 1980s and '90s, culminating in the National Organic Program standards issued in 2002, allowed for greater commensurability and increased the economic value of the organic market. In 1997, before the final USDA regulations for organic products were implemented, the organic industry was worth $3.6 billion (Dimitri and Oberholtzer 2009, iii). By 2000, the same year supermarkets became the primary source of organic food in the United States, it was worth $7.8 billion. In 2015, well after organic foods were fully regulated by the USDA and became a fixture in American supermarkets, the organic industry reported annual sales of over $39 billion (Dimitri and Greene 2002, 1; OTA 2015).

The contemporary prominence and mainstream acceptance of organic as a niche market finds its surest expression in the entry of big-box retailers into the organic sector. Ever since the spring of 2006, when Walmart announced its intention to expand organic food offerings in its stores across the nation, Americans have been granted unprecedented access to organic products. From boxed macaroni and cheese to baby spinach, Walmart and other mega retailers (such as Costco, Super Stop and Shop, and Whole Foods) provide consumers with the bulk of organic produce available.

The rise of industrial-scale organic agriculture and food production did not begin with Walmart. More accurately, it began as early as the 1990s, when the concept of organic was being stabilized by government regulation and when corporations such as Gerber, Heinz, Welch Foods, and even the Walt Disney Company began entering the growing organic food production market by buying out several smaller-scale organic companies (DeLind 2000; Murphy 1996; Pollan 2006, 154). The *New York Times* investigated these acquisitions as early as 1996, citing evidence that organic was increasingly "big business" and reporting that major retailers—like the A&P supermarket chain—were devoting increasing square footage to organic products (Murphy 1996). A wave of even more acquisitions would occur after the National Organic Standards were implemented in 2002 (Howard 2009). Corporations began producing organic food on an unprecedented industrial scale, and supermarkets began to sell organic products at seemingly ever-increasing volumes as the demand for organic products continued to grow.

Walmart's decision to enter the organic food market was of tremendous consequence given the retail giant's prominence as the number one grocery retailer in the United States, a position it first occupied in 2001, based on volume of sales as a result of the expansion of its supercenter locations (see Lichtenstein 2009, 135). Almost immediately following Walmart's announcement that it would begin offering new organic options for consumers, articles began appearing in major news outlets questioning the impact of its decision (Bhatnagar 2006; Brady 2006; Warner 2006). Likewise, on the heels of this decision came an expanding cultural conversation concerning the value of eating industrial organic food.

For decades, alternative agriculture activists had worked feverishly to convince the American public that there was no better food than organic. By and large, these efforts were successful. Around the time of Walmart's announcement, almost one-quarter of Americans reported purchasing organic products at least once a week (Cloud 2007, 1). However, only a year after this announcement, a dramatic shift occurred in the conversation surrounding organic foods. *Time Magazine* even published an article with the provocative title "Eating Better than Organic," in which the author argued, "For food purists, 'local' is the new 'organic,' the new ideal that promises healthier bodies and a healthier planet" (ibid.). As a result of the shift from organic farming to organic industry, a new, local expression of alternative agriculture was increasingly seen as necessary. This dramatic shift away from organic food as the paragon of responsible eating must be viewed in light of Walmart's entry into the market and that of fellow retailers following in its footsteps. Indeed, the conventionalizing image of organic has spurred many consumers toward a valorization of local food; studies have shown that "the shift in preferences [toward local food] . . . has been largely a result of consumers turning away from industrialization of organic agriculture (Adams and Salois 2010, 336). The same shift has been reflected in studies of New England consumers (Berlin, Lockeretz, and Bell 2009).

Perhaps more than anything else, Walmart's entry into the organic food market serves to reveal precisely what variety of organic agriculture is being cultivated across the country. As we have argued, organic agriculture in the United States has represented an alternative to industrial agriculture and a rejection of the status quo by the diverse activist groups throughout the movement's history. Today, many organic farmers still share these concerns and work to maintain alternative agricultural practices. However, the shelf-stable packages of organic foods that are offered by the supercenters and supermarkets across the country bear little resemblance to this countercultural approach to the food system. Rather, a great percentage of the organic movement has been absorbed into the dominant food system and, as a result, offers little promise of delivering the types of sustainability motivating calls for an alternative food system.

Instead, the types of organic products the majority of Americans now purchase in supermarkets could be classified as a co-opted, mainstream form of organic agriculture that has lost the ability to challenge the failings of our industrial food system. This transition did not occur through a legislative battle to suppress organic agriculture or a media smear campaign. Rather, the legislative battles surrounding the creation of the National Organic Program reflected an effort to redefine organic in rational, economic terms and incorporate organic products into the cultural mainstream. In the words of critical theorist Herbert Marcuse, the "liquidation of *two-dimensional* culture takes place not through the denial and rejection of the 'cultural values,' but through their wholesale incorporation into the established order, through their reproduction and display on a massive scale" (1964, 57, emphasis in original). As the USDA redefined organic agriculture as a niche market within American agribusiness, consumers tacitly supported such a redefinition through their daily practices. Organic farming was no longer rejected by mainstream society as the ostensible stuff of iconoclasts and hippies; in fact, organic products were being touted everywhere as a good alternative to conventional ones. Through buying, selling, and talking about big-box organic as though it were synonymous with sustainability, the alternative nature of those practices gradually eroded. The co-optation of organic production by the conventional food industry blurs the distinctions between industrial products and sustainable alternatives.

While organic farming was once considered an entirely alternative form of agriculture and while this is still the case on some small-scale farms, it is no longer the case for the entire organic sector. The products of organic agriculture are now represented on store shelves from midtown to Middle America—but not because holistic organic practices have become the norm. On the contrary, organic farming is now largely carried out by agribusiness using many of the same agricultural techniques utilized on conventional farms and with little concern for economic and social sustainability. Rather than offering solutions to the many problems of sustainability facing our food system, the products of such agribusinesses are increasingly recognized as representing a conformist vision and practice of organic food production (Adams and Salois 2010; Johnston, Biro, and MacKendrick 2009).

Organic farming's conformity to the agribusiness model became most apparent in the production of organic lettuce early on. By the early 2000s, a single corporation, Earthbound Farm, grew 80 percent of the organic lettuce sold in the United States (Pollan 2006, 138). The company's success attracted attention. In 2013, WhiteWave Foods—a former subsidiary of food processing giant Dean Foods with ownership of five different organic brands—acquired Earthbound Farm, an acquisition resulting in further consolidation in the organic sector. At purchase, Earthbound Farm still controlled

60 percent of the market share for all branded packaged salad greens (Hennessy 2013). The same trends can be seen in many other organic foods. For example, by the mid-2000s Stonyfield Farm considered including powdered milk from New Zealand in its yogurt to keep up with demand, and the company was forced to cut the percentage of its products that were actually organic (Brady 2006, 1). Meanwhile, Horizon Organic Dairy—the nation's top brand of organic milk— was managing a massive operation with 8,000 cows (3). Aurora Organic Dairy, which produces store-brand organic milk for several supermarket chains, kept up with demand with 8,400 cows kept on merely two "farms" (4). Due to the pastoral undertones associated with the word "farm," such operations, consisting of thousands of cows, are perhaps more aptly described as "industrial-organic bovine factories." After numerous boycotts and formal complaints to federal regulators, both companies have recently reduced the number of cows in their herds and expanded their animals' access to pasture. However, new complaints allege that Horizon has increased how often its animals are milked, from two times to as much as four times a day. Under this system of management, high-producing cows remain confined to milking parlors, without access to pasture, until the animals' milk production levels lag (Cornucopia News 2014). Such practices hardly challenge the standard conventional agricultural practices of our time. In fact, they are in many ways indistinguishable from them. And this only makes sense: less than a decade after the introduction of the USDA's organic standards, fourteen of the top twenty North American food processors had either acquired or introduced an organic brand (Howard 2009, 26–27).

Not only have the "new" organic products being offered to consumers conformed to conventional standards of industrial production, but they have also been produced for conventional food interests. For example, for several years before Walmart entered the organic industry, several of its top grocery competitors were offering an array of organic products. Corporations such as "Safeway, Kroger, and SuperValue . . . have private label organic lines with names like Nature's Best and O that they sell at prices below those of brand organic products" (Warner 2006, 1). Far from creating countercultural modes of food production, distribution, and consumption that could increase social connection and a more just economy, the organic industry has increasingly begun to replicate what it set out to contest.

More than just agribusiness as usual, the new organic products lining the shelves of supercenters represent an unobtrusive organic agriculture that allows consumers to ignore much of the movement's original mandate to improve conventionalized food systems. In order to fit on grocery store shelves, organic farming has been re-envisioned to complement existing food products rather than supplant them. Describing Walmart's decision to carry organic food, Bruce Peterson, in charge of Walmart's perishable food sales, went so far as to say that it "is like any other merchandising scheme we have, which is providing customers what they want" (cited in Warner 2006, 2).

The chief operating officer of Kellogg has gone even further in an effort to prevent "organic" from being anything more than a preference, rather than an alternative to the unsustainability of the contemporary food system. After Kellogg began producing organic Rice Krispies cereal, David Mackay, then chief operating officer, insisted, "We have no intent to send a message that the standard Rice Krispies are somehow not great brands" (cited in Warner 2006, 2). The rise of organic agriculture was, in large part, the result of a belief that the standard products of conventional agribusiness were, somehow, not all that great. However, on a quest for profit—not change—the new organic has become an unobtrusive companion to conventional food products.

The very fact that the chief operating officer of Kellogg is discussing organic Rice Krispies reveals another characteristic of the new organic products that have arisen in response to the full-force entry of corporations into the organic market. The new organic products are entirely palatable to mainstream American consumers because they are presented with the same products that conventional agribusiness has offered them for decades. While some corporations were planning on offering organic versions of their popular products before the spring of 2006, Walmart's entry into the organic market directly influenced others. According to the senior vice president for dry groceries at Walmart, the company had spent the preceding year urging its suppliers to offer organic products (Warner 2006, 1). Such an approach was predicted to be highly successful by Walmart's Bruce Peterson, as it would rely on existing brand loyalty and advertising budgets (Warner 2006, 2). Although organic agriculture once existed in a state of uncomfortable tension with the mainstream food industry, organic foods were now—more often than not—the comfort foods mass-produced by corporate agribusinesses.

The evidence for this conclusion lines the aisles of even the "crunchiest" natural foods stores. By 2008, "Organic manufacturers . . . were either competing directly with conventional food manufacturers or had been subsumed by conventional firms" (Dimitri and Oberholtzer 2009, 1). Perhaps even more troubling is the fact that "in the U.S. General Mills owns several large organic brands including Muir Glen (organic processing tomatoes) and Cascadian Farms (organic frozen fruits and vegetables), and markets its own organic breakfast cereal, Sunrise. Gerber, Kellogg, Mars, Heinz, and Dole own or sell at least one organic product" (Guthman 2004c, 304). And the acquisitions continued: by 2015 Hormel had acquired Applegate Farms and WhiteWave Foods had acquired Wallaby Yogurt (Howard 2015). Even Annie's—the makers of boxed macaroni and cheese with a cultish following among "harried, organo-hipster parents everywhere" (Marx de Salcedo 2007)—is now just a subsidiary of General Mills (Howard 2015). On supermarket shelves filled with mass-produced foods, it seems as though few organic products stand out as genuine alternatives to the conventional food system. Even those that appear to be are, more often than not, just cleverly disguised.

Philip Howard has done much to reveal these often subtle changes on the supermarket shelves. Howard's research shows that the movement of agribusiness and multinational corporations into the organic food market has failed to rouse consumers attracted to its countercultural roots to action since "the process of conventionalization

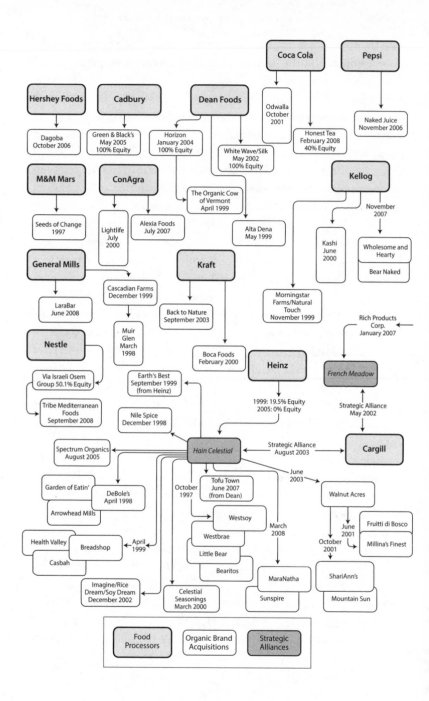

has been well-hidden from them through practices such as 'stealth' mergers and acquisitions" (Howard 2009, 14–15). As a result, the products lining the shelves of natural food stores continue to appear as if they are legitimate alternatives to the conventional food system. However, this is merely an illusion. For example, and further illustrated in Figure 1, Howard reveals that "most General Mills brands have the company's traditional 'G' logo on their packaging, but Muir Glen and Cascadian Farm products do not" (18). Clearly, the conventional food system has managed to co-opt most of the products that line the shelves of retailers, but they have also tried to keep this process quiet.

Just as they have co-opted many of the products of the organic food movement, these corporations have also managed to co-opt much of the organic movement's message. For example, Howard points out that "Horizon, which has been boycotted by a number of retailers for its nominally organic feed-lot dairy practices, for instance, depicts a smiling cartoon cow in front of the planet Earth on its milk cartons" (Howard 2009, 18). Large-scale retailers involved in the organic industry have also done their part keeping up appearances. Michael Pollan has likened this to the creation of a new literary genre: "Supermarket Pastoral" (2006, 134–139). Describing an experience shopping at Whole Foods (identified by Pollan as the supercenter offering the most "cutting-edge grocery lit."), Pollan recounts the stories he is told about his food in an attempt to link the consumer of industrial organics with ideals of sustainability: "eggs 'from cage-free vegetarian hens'" and "milk from cows that live 'free from unnecessary fear and distress'" (135).

These environmentally friendly pastoral images sell consumers a bill of goods. More disturbing, however, is how deceptive these messages

Figure 1. Industry Structure of "Supermarket Organic": Major food processors' acquisitions of organic brands. (Adapted from P. Howard, "Organic Industry Structure: Acquisitions by the Top 30 Food Processors in North America." https://www.msu.edu/~howardp/organicindustry .html. Accessed September 20, 2009.)

can be for consumers. For example, Bear Naked, an organic granola brand, presents an abridged time-line of the company's history on its website, "beginning with the two founders meeting at age 11, then tracing their initial $7,000 self-financing and growth (supplemented by 0% interest credit cards), *but ending just before their sale to Kellogg*" (Howard 2009, 19, our emphasis). With empirical evidence for the co-opting of the organic products lining grocery store shelves, there may be little "alternative" about the forms of organic agriculture carried out by industrial food growers and manufacturers.

The incorporation of organic foods into the established order has not only been occurring among food manufacturers. Rather, much of the farming of organic food in fields across America has come to appear remarkably similar to conventional agribusiness. The question of the extent of conventionalization in the organic sector has been given significant attention by agro-food scholars (e.g., DuPuis and Gillon 2008; Guthman 2004a; Jaffee and Howard 2010; Obach 2007). For example, in her research in California, Guthman found that "few organic growers see organic farming as a means to alternative institution building, although many are explicitly distrustful of the worst of conventional agriculture. More significant perhaps, even the dyed-in-the-wool all-organic movement growers are becoming less ideologically radical and are adopting practices they might have otherwise shunned" (2004a, 60). Guthman argues that "a focus on allowable inputs has minimized the importance of agroecology, enforcement has become self-protecting and uneven, and reliance on incentive based regulation has profoundly shaped who can participate and on what terms" (111). The presence of corporations as competitors to small organic growers often prevents committed farmers from remaining true to their ideals (178).

In a similar vein, Jaffe and Howard argue that the "bar lowering" that resulted from a regulatory focus on allowable inputs "has clearly reduced the internalization of costs" (2009, 394), most often externalized by conventional agriculture; costs shifted onto farm workers, rural communities, and the environment. As they note, "What remains . . . is, increasingly, a single meaningful variable: the absence of synthetic agrochemicals," making it easier for industrial actors to participate in organic and place organic options on supermarket shelves. The pressures of conventionalization exert further influence on the organic market through price competition. As Michael Bell explains, the competition wrought by the treadmill of production in capitalist agriculture is particularly intense. "The inelasticity of demand for food," he explains, "combined with the regional specificity of most agricultural production and the regional crises that weather often brings, means that farmers experience their own treadmill in a particularly intense and local way" (2004, 42). Bell calls this unique condition "the farmer's problem." The result is that, as a farmer, "you pretty much only make money when you have a good year and a lot of other similar farmers don't. And that doesn't happen very often" (43). The problem of competition and the erosion of price premiums is even more intense in a market such as organics: to be sure, there is an "organic farmer's problem."

In the context of California, Guthman has argued, "a regulatory structure that only attempts to support a price premium is exactly what contributes to the erosion of organic practices . . . through rampant competition" (2004c, 313). Economic premiums are prone to erosion through competition. As a result, in an agricultural market like organics, where producers are dependent on price premiums as compensation for undertaking less profitable ecological farming practices, the erosion of premiums through competition (particularly competition with larger corporate farms benefiting from economies of scale) inevitably yields pressures driving farms toward more conventional practices. This focus on competition over price, however, cannot be limited to California since it has produced most of the country's fresh organic produce for many years, totaling about 60 percent of organic vegetables and 50 percent of organic fruit in the United States during the early 2000s (C. Greene and Kremen 2003). According to the USDA's Organic Survey conducted in 2014, California still comfortably leads the nation in organic production, accounting for 40 percent of all U.S. organic sales (Young 2015). As a result, small organic farms from the Midwest to New England are forced to compete with supermarket organic—largely supplied by California agribusinesses.

Such changes within the organic market lead naturally to questions about whether the now dominant form of "supermarket organic" food is capable of addressing problems of sustainability. Given that the mainstream organic movement has focused disproportionately on issues of environmental sustainability, we too will begin with that concern. Faced with the explosive growth of the organic sector, one might reasonably ask, "What is wrong with supermarket organic?" So what if organic agriculture has become increasingly enmeshed in the conventional food system—if it has favored economic rationality to pursue a strategy of growth? Is it not better that organic practices are expanded to an industrial scale, transferring countless acres of farmland to organic production and removing countless tons of pesticides, herbicides, and synthetic fertilizers from the environment? The organic movement is merely benefiting from the efficiency of the larger food system, thereby extending its reach.

The Problem with Supermarket Organic

There is, of course, something to the seemingly utilitarian philosophy of the argument. For example, if we consider all of the farms that contract to produce Earthbound organic salad greens, we see that twenty-five thousand acres of land are being farmed with USDA-approved organic methods. It is estimated that this production represents the elimination of "270,000 pounds of pesticide and 8 million pounds of petrochemical fertilizer that would otherwise have been applied, a boon to both the environment and the people who work in those fields" (Pollan 2006, 164). Yet the expanding corporate control of the organic sector exerts tremendous pressure toward both the further fragmentation (i.e., increasing specialization in one aspect of food production) and the homogenization (i.e., the capture of market share by national, often global, brands) of the food system (Gillespie et al. 2007, 67). Such pressures have tremendous consequences for the environmental sustainability of today's farming operations.

Specifically, these pressures erode environmental sustainability by leading farmers to specialize their operations in order to produce at the scale a corporate food system requires. To compete in a fragmented and homogenized global food system, farmers must "emphasize economic efficiency and quantity of production rather than social values and product quality" and reduce the "agroecological complexity in any given farming operation" (Gillespie et al. 2007, 68). While the reduction in pesticide use even on industrial organic farms may be laudable, the realities of producing food for national or global distribution limits the ability of farmers to employ the types of agro-ecological approaches to managing their land necessary for limiting the environmental impacts of contemporary farming. As a result, the organic foods that fill most produce aisles of contemporary supermarkets—and the production systems that have emerged to supply organic foods at a supermarket-sized scale—can barely make a dent in the agro-ecological concerns about the environmental issues faced by contemporary agriculture (P. Allen and Kovach 2000; Guthman 2004a, 2004c). According to Julie Guthman, the farming practices that have brought us to this contemporary era in the organic movement's history, in which the supermarket is the most common place consumers come in contact with the word "organic," are hardly sustainable (1998, 137): what they offer is an "organic-lite" (2004c, 301).

These types of organic production methods and social relations are even less likely to improve the social and economic sustainability of the U.S. food system. For example, the rise of industrial agriculture in the United States in general has had tremendous consequences for the economic and social resiliency of local communities (Gillespie et al. 2007; Lyson and Green 1999). Much of industrial-scale agriculture's efficiency comes from the same fragmentation and homogenization of the food system, and the resultant consolidation of control and intensification of production, that we described in relation to industrial organic's inability to address environmental sustainability. It is the "get big or get out" concept of agriculture popularized by Earl Butz. As a result, since the end of World War II,

"America has lost about one farm every half hour" (McKibben 2007, 54).

As a result of these requirements for producing food on an industrial scale, much of the social and economic infrastructure of farming communities has been similarly homogenized and fragmented (Lyson and Green 1999). For example, in the 1990s, the loss of population caused by the decline of agriculture became so great that "an area the size of the original Louisiana Purchase again qualifies for the 'frontier' designation that the Census Bureau gave remote regions before the great waves of settlement in the nineteenth century" (Nichols 2003). For the farmers remaining in the industrial food system in these communities, the economic returns provided are "extremely low" (Lyson and Green 1999). While poverty is often cast as an urban issue, "nine of the ten counties in America with the lowest per capita income are in farm states west of the Mississippi" (McKibben 2007, 57). Wondering how politicians allow this process to go on, Helen Waller, a farmer in McCone County, Montana, remarked, "I wonder if it's because they've gotten so used to measuring everything in economic terms that they don't recognize that behind all these numbers from all these forgotten places, there are people who are hurting" (cited in Nichols 2003).

In contrast, small-scale, local production can be a very efficient engine for local economic growth. For example, studies have shown that small farms can be more efficient in terms of "converting inputs into outputs," producing more food per acre (Halweil 2004, 76). Moreover, the benefits of that productivity are more likely to stay in the local community. In 2008, only 11.6 cents of every food dollar spent made its way back to American farmers (Canning 2011). Most of each dollar is spent outside of the farmer's community. However, a study in Vermont found that "if local consumers 'substituted local production for only 10 percent of the food we import, it would result in $376 million in new economic output, including $69 million in personal earnings from 3,616 new jobs'" (cited in McKibben 2007, 165).

For some, the consolidation of agriculture brought about by industrial fragmentation and homogenization is of little consequence.

According to the agricultural economist Stephen Blank, "Most Americans could not care less if farming and ranching disappear, as long as they get their burgers and fries. America will waddle on" (Blank 1999, 33). For Blank and many others, farming is a part of America's past—not its future. Yet studies have consistently provided evidence that the rise of industrial agricultural operations has reduced the social capital and economic vitality of the communities surrounding them, relative to more local systems of production, ownership, and control (for a review, see Lyson and Green 1999). In contrast to industrial production, local systems of production have been found to be reliant upon diverse social and economic infrastructure, serve as catalysts for entrepreneurial activities, and facilitate greater non-market social interactions alongside economic exchanges (Chiffoleau 2009; Gillespie et al. 2007). Given the fact that the organic food consumers encounter at large grocery retailers is supplied by the same industrial food supply chain as the conventional food alongside it, it cannot possibly contribute to rebuilding the social and economic infrastructure the industrial system of agriculture has eroded in local communities.

"Supermarket organic" is also limited in its ability to foster social connections and create spaces for civic participation in the food system. In fact, the organic label emblazoned on products may actually stifle such engagement. Organic certification was implemented to serve as a rational, commensurable proxy for "engaged consumption," allowing consumers to "assume authenticity, rather than identify it through engaged interaction" (Mount 2012). The label, much like a brand, serves as a commensurable means of distinction: an economic tool to facilitate ideal market exchange. Despite the fact that increased social connection is one of the most promising aspects of alternative food systems (Feegan and Morris 2009; Gillespie et al. 2007), the supermarket setting in which most organic food is purchased limits the potential of "organic"—in and of itself—to bring us closer to more sustainable social and civic relationships. Studies of relationships in various venues for food purchases show that supermarkets foster the least rich, most perfunctory interactions among patrons when compared to neighborhood grocers' and farmers' markets (Cicatiello et al. 2015), a pattern we observed in our own observations of the spaces in New England in which organic food is sold.

Moreover, a sustainable food system must promote a civic-minded approach to agriculture and food provision (Trauger et al. 2010) and foster spaces of citizen engagement and public discourse (Gillespie et al. 2007). Ideally, a sustainable food system would provide a "food democracy" where food would exist "within democratic control in the commons" (Johnston, Biro, and MacKendrick 2009, 524–525). In the supermarket, "what the corporate-organic foodscape arguably provides is thus a *simulation* of place, locality, and humanized producers: images of precisely those things that are destroyed by capitalism's tendency to subsume everything to the law of value" (525). And yet, such spaces of civic engagement and social connection are desperately needed—they have been in decline across various facets of contemporary American life for decades (McPherson, Smith-Lovin, and Brashears 2006; Oldenberg 1989; Putnam 2000). Outside of the grocery store, alternative local food systems show promise of fostering precisely these types of civic spaces and community (Cicatiello et al. 2015; Feegan and Morris 2009; Gillespie et al. 2007; Trauger et al. 2010). Such spaces of sociability and civic engagement can be particularly important places for grassroots organizing in places where the effects of racial and ethnic inequality render neighborhoods and communities "otherwise stark environment[s]" (Alkon 2008).

It's Not Just the Supermarket: The Social Problems USDA Organic Can't Fix

The problems of the contemporary organic movement are not limited to the fact that more consumers are making their organic purchases at the supermarket checkout. Even in more alternative spaces, the narrow focus of contemporary ideas about "organic" leaves many social problems in the contemporary U.S. food system unchallenged—or even exacerbated. While the regulation of the organic industry has led to tremendous economic growth, benefiting corporate actors and elevating farm owners to often celebrity status, the continued plight of farmworkers has largely remained unaddressed by the emergence of the contemporary organic industry and by much of the contemporary organic movement (P. Allen 2004; Harrison 2011). Indeed, it has remained largely unconsidered. Consumers at farmers' markets—let alone in the hectic post-work rush in the supermarket—have been found to rarely think about farmworkers, reproducing a "white farm imaginary" in which sustainable agriculture is done by "white family farmers" (Alkon and McCullen 2011, 946). This narrative "[renders] farmworkers invisible" and leads to incorrect assumptions that organic farmers "hire only relatives, making regulation of employment practices seem unnecessary" (947). Far from problematizing the notion of "an economy based upon farm owners who hire landless laborers to plant, tend, and harvest crops," when they are discussed at all in the predominant discourses of the contemporary organic movement, farmworkers are often cast as just another economic input "along with equipment and fuel" (P. Allen 2004, 136).

Outside of neglecting concerns about the welfare of our food system's farmworkers, the contemporary organic sector does little to redress the social inequalities produced and exacerbated by our contemporary food system. By and large, the contemporary organic movement has been criticized for perpetuating race-based inequalities through performances of whiteness (Guthman 2011; Alkon and McCullen 2011). Related to the issue of the invisibility of farmworkers, "by focusing on and heroicizing farm owners, rather than farmworkers, the alternative agriculture movement emphasizes and valorizes the role of whites in the food system rather than people of color" (Alkon and McCullen 2011, 947).

Moreover, the very framing of supermarket organic as empowering consumer choice—much the way the executives at Walmart and Kellogg describe it—obscures the racial and class privilege involved in purchasing organic foods (Guthman 2008a; Johnston and Szabo 2011). The contemporary food system has been predicated on a system of white privilege in which "land was virtually given away to whites at the same time that reconstruction failed in the South, Native

American lands were appropriated, Chinese and Japanese were precluded from land ownership, and the Spanish-speaking Californios were disenfranchised of their ranches. Given this history, it is certainly conceivable that for some people knowing where your food comes from and paying the full cost would not have the same aesthetic appeal that it does for white, middle-class alternative food aficionados" (Guthman 2008a, 394). Relatedly, the vision of organic food as a consumer choice cannot redress the class-based inequalities of our currently unsustainable food system. Many consumers have been found to be utterly unreflexive concerning the class privilege underpinning the ability to choose an organic option (Johnston and Szabo 2011), and organic foods have become an entrenched marker of class status and foodie distinction (Guthman 2003; Johnston and Baumann 2010).

The framing of organic food specifically, and green consumerism in general, also intersects with gender to reproduce, rather than challenge, the inequalities of the contemporary food system. While women farmers have challenged hegemonic gender identities through their participation in alternative agriculture (Trauger et al. 2010), the burden of ethical green consumption has primarily fallen on women (Sandilands 1993). Women have long been identified as being more likely to consume organic foods, often as a result of their greater likelihood to engage in household provisioning and the care of children (Lockie et al. 2002). As a result, the contemporary organic movement's emphasis on "shopping for change" (Johnston and Szabo 2011) means that "to be 'environmentally friendly,' to be 'kinder to nature,' *requires* the use of "green" products, *requires* recycling, *requires* the intensification of household labour" (Sandilands 1993, 47, emphasis in original)—additional labor that disproportionately falls upon women.

The overwhelming consensus is that the majority of today's organic food going through the supermarket checkout line, and even some of the food changing hands in farmers' market stalls, cannot possibly deliver on the broad goals of sustainability. The consolidation and conventionalization of the organic industry, spurred by the tendencies toward fragmentation and homogenization on which the modern industrial food system is predicated, preclude many of the characteristics an environmentally, economically, and socially just and sustainable food system would require. While the regulation of organic farming over the past three decades allowed organic agriculture to expand by rationalizing practices and ensuring consistent standards of commensurability, organic farming at its nascence in the modern world was never intended to rely solely on economic rationality. Throughout its various ideological permutations, the historic organic movement was predicated on recognizing multiple rationalities to produce economic, environmental, and social value. Organic farming was meant to be smaller in order to be agro-ecological, community-based, and community building (Altieri 1995). More important, many farmers today still hold to these values.

Spaces for Alternatives in a Bifurcated Market

Not all organic farms are industrialized or operated as "chemical-free" counterparts to conventional farms, and farmers, farmworkers, consumers, and communities are working toward viable and more expansively sustainable solutions to the problems of our industrial food system—conventional and organic. As outlined in the introduction, far from uniformly conventionalizing, as the direst accounts of the contemporary organic sector predicted, the organic market has bifurcated. Specifically, bifurcation is described as "the process by which . . . organic agriculture adopts a dual-structure of smaller, lifestyle-oriented producers and larger, industrial-scale producers" (Constance, Choi, and Lyke-Ho-Gland 2008). The bifurcation approach has suggested that the unique ecological constraints of farming (such as its seasonality), the relative productivity of small farms in certain circumstances, and political concerns about the safety and quality of industrial foods provide niches for small-scale organic production (Coombes and Campbell 1998, 141).

Often, these small-scale organic operations are predicated on the development of local food networks and more relational forms of economic exchange directly with consumers (Brehm and Eisenhauer 2008; Chiffoleau 2009; Feenstra 2002; Gillespie et al. 2007; Seyfang 2006). While not addressing bifurcation explicitly, Brian Donahue portrays its impacts—and the ways a bifurcated organic market carves out potential spaces for more alternative forms of agriculture—in the New England context. Donahue argues, "The superiority of fresh local fruits and vegetables is compelling even in today's cheap energy economy, and the demand for organic food steadily grows. Such produce can be marketed quite competitively by small, intensive operations, even as small as five acres or less. Many pieces of land in the interstices of the spreading suburbs, too small to be farmed efficiently by modern, conventional means, can grow organic produce at a profit" (1999, 100). Particularly in many parts of New England, a region with a geography often ill suited to agro-industrial cultivation (see Bell 1989, 465), small-scale organic operations can have a competitive advantage. Donahue concedes that Land's Sake—a community farm in the Boston suburb of Weston—might be able to double its revenue if it were farmed organically in a more intensive, conventionalized manner. However, he also alludes to another reason small farms can continue to succeed organically: these farms can provide something intensive organic farms can't. They can draw consumers to the land (90).

It is important that while there is broad consensus that the types of organic products supplied by industrial food supply chains to supermarkets nationwide will not provide a holistically sustainable food system, there is also evidence that approaches that create alternative social relations in the food system—alternative ways of growing, exchanging, and connecting over food—can be useful starting points for transforming the contemporary agricultural economy. Farmers' markets, for example, have been pointed to as a "keystone" institution that can help to encourage the types of on-farm diversity that sustain more agro-ecological forms of farming, rebuild local economies, and serve an important role in expanding entrepreneurial opportunities, public civic space, and social connection (Gillespie et al. 2007).

This is not to say that local food initiatives are always unproblem-
atic solutions to the problems of agricultural sustainability (Born and
Purcell 2006; DuPuis and Goodman 2005; Hinrichs and Allen 2008).
We should remember that the reproduction of white privilege and
class status happens not only in the produce aisles of Whole Foods
(Johnston and Szabo 2011), but also among the shoppers and vendors
of California farmers' markets (Alkon and McCullen 2011). However,
as the literature on the political economy of contemporary organic
agriculture would suggest, such local niches within a bifurcated organic market provide structural space for farmers, communities, and
consumers seeking an alternative, organic future.

While such structural holes provide hope for spaces of sustainable
food practice, a relational approach to the economic exchanges in the
contemporary organic sector can attend to the ways farmers take advantage of these spaces outside the realm of their economic considerations. Left unexplored in approaches that identify such structural
possibilities for alternative farming are the ways small-scale farmers
engage in relational work to make sense of their place in the organic
landscape, how they make decisions in light of very real economic pressures and equally real ideological commitments, and how sustainable
and meaningful organic practices emerge in their fields.

The remainder of this book can be viewed as an exploration into
how small-scale organic farmers understand such realities. We know
that a sizable acreage in New England is put toward organic agriculture and that the average farm size is quite small (see figure 2). This
makes New England a unique place to observe how bifurcation plays
out in practice and how small-scale organic farmers determine and
make sense of their own place in the organic sector. New England's
unique history and geography have allowed it to be one of the most
historically productive agricultural regions in the United States; yet
such conditions also have rendered it unsuitable for the "highly mechanized, big is beautiful" model of agro-industry (Bell 1989, 465). As
a result, it may not be surprising that New England contains many
vibrant, small-scale organic farms. If California is the epicenter of

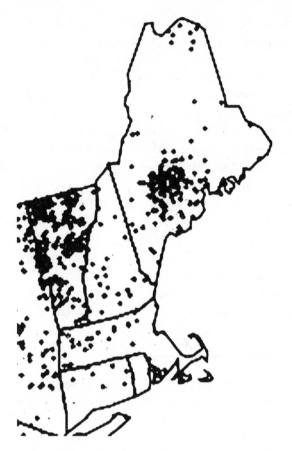

Figure 2. Acres Used for Organic Production in New England, 2007:
Distribution and concentration of land used for organic agriculture,
with each dot representing 250 acres. (Adapted from USDA, "Acres
Used for Organic Production: 2007," 2012. 07-M097. 2007 Census of
Agriculture. National Agricultural Statistics Service. http://www
.agcensus.usda.gov/Publications/2007/Online_Highlights/Ag_Atlas
_Maps/Farms/Land_in_Farms_and_Land_Use/07-M097-RGBDot1
-largetext.pdf. Accessed July 15, 2015.)

conventionalization (Guthman 2004a), New England may be an excellent counterpoint. However, given the focus of both the conventionalization and the bifurcation approaches on the political economy of organic agriculture, we know surprisingly little about how such farmers—situated outside the conventionalized organic mainstream—both understand and envisage organic futures on their farms and in their lives.

And yet enacting such an organic future does not happen in a vacuum, free from encumbrances. While the organic market has bifurcated and niches allow some farms to avoid conventionalization, farmers still face substantial pressures as agro-industry pushes out more cheap organic food than ever before. To envisage an organic future, small-scale farmers must work to forge new systems of social relations in an era characterized by supermarket organic. As we have noted above, these farmers must make what economic sociologists have called good matches, coming to workable understandings and practices given often contradictory needs and commitments. How do everyday farmers make the hardships of small-scale organic farming fit into their experiences of daily life? How do they forge a match between their livelihoods and the other social aspects of their lives? And how close can these matches get them to the organic ideals that agricultural scholars would suggest are necessary to achieve truly sustainable farming systems?

In the pages that follow, we provide a firsthand account of a New England organic farm, supplemented with stories and interviews from other small farms located mainly in New England but also in other parts of the United States. Driven by our commitment to push beyond an understanding of the political economy of organic to an understanding of how sustainable practices are enacted in a bifurcated organic market, we felt it important to gain access to the subjective experiences of organic farmers to determine what organic farming meant to them. Even more important, given our theoretical approach, was our ability to observe good matches where they were made: in the fields of a small New England farm. We present the stories of farmers whose practices genuinely reflect their deeply personal beliefs and values surrounding "organic" in the agro-ecological sense. However, it is also a realistic representation of their struggles, revealing the forces that constrain these individuals' actions, and those of all organic farmers in the United States and likely beyond. In doing so, we do not endeavor to depict the struggles of every small-scale organic farm in a generalizable manner. Instead, we seek to depict the mechanisms of good matches and relational work as explanatory tools, allowing us to understand how organic farmers negotiate the uniquely situational tensions they face while cultivating their organic futures.

PART II

The Land

PRELUDE: A SENSE OF PLACE

The conventionalization of the organic sector has led many to be skeptical about the possibility of organic in and of itself promoting a meaningfully sustainable alternative to the modern agricultural paradigm. However, the bifurcation of the organic market into both mainstream and alternative spheres provides places within the food system where investigating the work involved in creating new economic and social relations of farming practice makes sense. We may very well have come to an era in which most organic products pass through the same types of industrial commodity chains as organic initially set out to challenge. But more and more people desire a new type of economy, one that fosters social connection, meaning, and new relationships of exchange (Schor 2010; Schor and Thompson 2014). And more and more people are willing to buy into this new alternative model.

Many of the very qualities that those operating within spaces of alternative food production and consumption endeavor to attach to foods—"local" or "authentic," as examples—contest the narrow rationality of the market. In contrast to the market's ability to deliver standardized, uniform, commensurable, and interchangeable products, these types of alternative values demand particularity, place-specific meanings, personal connections, and important stories (Pratt 2008). As a result, small-scale farmers, specialty producers, and their consumers are actively constructing "the local" and its associated social relations through "innovation and contestation" over places, their products, and their meanings (Weiss 2011, 456). While such efforts can often fail to produce the types of inclusive sustainability our food system requires to better serve people and the planet (Johnston and Baumann 2010; Pratt 2008; Weiss 2011), at the very least, as sites of contestation, they represent spaces of alternative possibility.

Since social movement challenges to the dominant market rational-ity are often organized around these types of alternative cultural values (Pratt 2009, 172; see also Rosin and Campbell 2009), the re-lational approach to farmers' economic lives we take in this book is particularly appropriate. As social movements increasingly concen-trate their efforts on the market, it has become more important than ever to understand how cultural values can influence economic deci-sions in ways that either foster or stifle opportunities for meaning-ful social change (Schurman and Munro 2009). Ethnographic case studies do important theoretical work because they allow us to ap-proach efforts at creating alternative forms of agriculture "in terms of their own objectives and practices." As a result, they help us to avoid "naturalizing and universalizing" the role of capitalism's "rational-ity of maximizing monetary returns" in the lives of the very farmers who are often endeavoring to resist such visions of the "rational" mar-ket (Pratt 2009, 172). At the same time, such an approach allows us to reveal the ways the pressures and regulation of the broader market map onto individuals' lives and work (Grasseni 2003).

Ethnographies of small farms and farming communities have re-vealed the importance of the social relations of exchange, cultural meanings, and values in shaping the economic exchanges of food sys-tems. A particularly striking example is found in southwestern France, where a history of affordable land has led to an abundance of alternative, small-scale, and artisanal producers (Gowan and Slocum 2014). Gowan and Slocum's ethnography of the region demonstrates the importance of favorable socioeconomic conditions for the suc-cess of alternative economic practices of agriculture. Yet their find-ings are further reaching. They describe the ways individuals must negotiate a balance between their work and personal values about the nature of a life well lived, as well as the critical role of social con-nections in sustaining their alternative practices. They find that "the prevalence of gifting, cooperation, and informal exchange" within the region "is key to the high quality of life and economic survival" of small, alternative producers (31).

Among Italian Alpine cheese makers, European-origin labeling regulations have created markets for typical regional products. These market conditions have led producers to forge relationships with scientific and technical advisers, who have promoted standardized cheese cultures among artisanal producers to allow for more consistent local products in an increasingly global market. To exploit market demand and influenced by these advisers, these cheese producers have made a good match among their practices, beliefs about consumer desires, and economic interests. However, it has come with the increased reliance of a regional production system on the dynamics of the world market and on industrial suppliers of biological cheese cultures (Grasseni 2011). It bears noting that not all good matches are "good" or sustainable. Rather than representing a normative assessment, the concept of good matches merely recognizes how economic actors resolve contradictions between their market activities and a host of other social, ethical, and emotional commitments to arrive at viable economic arrangements and practices (Zelizer 2010). However, ethnographic studies of these same Italian cheese producers have shown that some of their relationships and values produce sustainable market practices. For example, many farmers have worked to reintroduce native livestock breeds, often through their relationships with technical advisers who monitor these "new" animals' productivity. These more alternative practices are further supported by farmers' connections to activist organizations like Slow Food, which help promote their work, and through "increasing dependence on wider networks of carefully targeted quality consumers" (Grasseni 2014, 65).

Similar connections are important in the market relationships of those involved with local, pasture-raised, heritage-breed pigs in North Carolina, where "in the shadows" of the explosive growth of industrial hog farming, a coalition of "farmers, restaurateurs, and consumers" have been co-constructing an alternative set of social relations for the production and consumption of local pork (Weiss 2011, 439). Through such relationships, and reflecting social values about how local pastured pork should taste, these networks have produced a market for a unique breed of pig (the Ossabaw Cross) and a unique set of production practices (pasture raising, often with particular types of forage). Restaurants in the area include local Ossabaw pork on their menus, even when they know that the expensive production practices mean they rarely can break even on their own cost of procuring it. In doing so, they promote interest in the locally grown product and help sustain an alternative set of economic relationships around food that support agricultural practices offering far greater prospects of long-term sustainability than the industrial hog-farming system that still dominates North Carolinian agriculture (Weiss 2011).

Implicitly, these accounts (and others noted in this book) suggest the importance of understanding the relational work involved in creating alternative food markets. Approaching our ethnography of Scenic View Farm, the case study at the heart of this book, from the perspective of relational economic sociology, we make such connections explicit. In doing so, we offer a systematic framework for understanding how particular forms of relational work create spaces for alternative practices in farmers' lives, contributing to the modern bifurcation of the organic sector. More generally, we show the ways actors engage in relational work to extricate economic relations from the domain of the dominant "rational" market and attempt to build alternative, more sustainable, and more socially connected visions of contemporary economic life. This approach means that we focus on the farmer but also consider seriously the other actors involved in making these alternative networks possible—the workers, the consumer-participants, and the land.

We have already briefly introduced Scenic View Farm, and in the pages that follow our main focus will be the ways the people who work and are connected to the land there navigate these issues. This story's strength lies in its attention to the details, uncovering the ways each of these farmers and farmworkers understand and practice their life's work. In order to engage in the sort of "fine-grained assessment" of sustainable agriculture needed to truly understand what farmers are trying to accomplish through their alternative practices (Pratt 2009, 172), it is equally necessary to first situate Scenic View in its particular context.

As the conventionalization thesis and theories of bifurcation so clearly demonstrate, organic farming practices take place within the wider political economy of contemporary agriculture (e.g., Buck, Getz, and Guthman 1997; Constance, Choi, and Lyke-Ho-Gland 2008; Coombes and Campbell 1998; Guthman 2004c). Scenic View exists within the larger world of organic agriculture in the Northeast and in the organic sector of the United States in general. Therefore, the case study draws from the complex history of the organic movement we have described in the previous chapters; the particularities of our case are instantiations of the acceptance of and resistance to the broader forces of the market, as this set of farmers and supporters engage in practices that attempt to make both their business and lifestyles viable.

New England, of course, contains a distinctive set of farming practices, and farmers have unique regional relationships with states, towns, and communities. Despite common misconceptions that New England was poorly suited to agriculture and that serious farming was abandoned in favor of more productive regions in the distant past, the region was once an agricultural powerhouse. Michael Bell notes, "In 1879, the first crop year for which per-acre yield information is available for the entire country, New England exceeded the national average for corn by 19 percent, for oats by 22 percent, for wheat by 16 percent, for barley by 7 percent, and for buckwheat by 21 percent" (1989, 456–457). Even to this day, by some measures New England's farms remain highly productive, and pockets of fertile, tillable soil remain despite suburbanization (Donahue 1999; Jager 2004).

Farming interspersed amid cities and towns has long characterized the farming of New England (Russell 1976). Just over one hundred years ago, Boston and the manufacturing centers surrounding it were supplied with fresh produce by a network of thousands of market gardens (Donahue 1999). In 1910, there were forty thousand acres of land in Massachusetts devoted to the market gardening of fruits and vegetables (101). Such forms of farming began a steady decline in the postwar era, their profitability undercut by the ascendance of industrial agriculture in the west and their land increasingly consumed by suburban sprawl (66–67). Nevertheless, the same model of agriculture—characterized by smaller farms located close to consumers—remains profitable in the region.

In New Hampshire, for example, Ronald Jager (2004) has reported that the bifurcation of the agricultural sector was showing promise—at least in terms of small-farm survival. New Hampshire's farms are largely divided between larger commodity farms and smaller niche-market farms. Of the approximately three thousand farms in the state, Jager reports that only four hundred could be classified as commodity farms. Meanwhile, the number of niche- and retail-farming operations was rising (50–51). Considering New England more broadly, Brian Donahue (1999, 74) has suggested—with evidence from the successful Land's Sake Farm in Weston, Massachusetts—that the market-garden model can still be profitable even with the pressures of suburban sprawl.

New England is also distinctive in the relative absence of minorities engaged in agriculture. Although the number of nonwhite farm operators is increasing nationwide, across the United States farm operators remain overwhelmingly white. The 2012 Census of Agriculture showed a 12 percent increase in black farm operators and a 21 percent increase for Hispanic and Asian operators. Nevertheless, black farm operators represent only 1.4 percent of U.S. farm operators, while Hispanics represent 3.1 percent and Asians merely 0.7 percent. However, these minority farm operators are highly concentrated regionally. For example, 90 percent of black farmers in the country are concentrated in twelve southern states (USDA 2014).

As a result, agriculture in New England is even whiter than in the rest of the country. In Vermont in 2012, for example, only 35 of the state's twelve thousand farm operators were black. Similarly, only 35 of them were Asian, and roughly 100 identified as Hispanic. Maine had only 94 black farm operators (or approximately 0.7 percent of the state's total), 18 Asian farm operators (about 0.1 percent), and 134

Hispanic operators (about 1 percent) in 2012. In Massachusetts, farm operators are slightly more diverse than in other New England states: in 2012, 0.9 percent were black, 1.2 percent were Asian, and 1.7 percent were Hispanic (USDA 2014).

Outside of farm operators, the labor markets of New England also suggest the whiteness of the region's agricultural sector relative to the rest of the United States. The USDA Census of Agriculture tracks the number of farms hiring migrant farmworkers—who, today, are largely of Hispanic origin (Arcury and Quandt 2009). Even in agricultural regions reliant upon their labor, migrant workers are often systematically unnoticed by residents, public officials, and consumers alike. Their employment is a "public secret," their presence hidden from, or avoided by, the diverted gaze of those more privileged (Holmes 2007, 2013). New England farms do hire migrant farmworkers—particularly for harvesting fruits like apples and blueberries—and their presence in the region is rarely remarked on until particularly deplorable labor conditions rouse the local media's attention (e.g., Blanding 2002). However, out of 7,755 farms in Massachusetts, only 132 hired migrant farmworkers (USDA 2014). As a result, the individuals we will describe who work the land at Scenic View Farm and other farms like it are representative of the racial privilege of the region's farm operators and the labor practices of many of the region's smaller farms.

New England's unique agricultural economy provides context for the good matches being made on Scenic View Farm. However, some of the pressures facing small farms in New England are wrapped up in general pressures affecting small-scale farming across the United States. Consequently, an understanding of some of the recent transformations of the American agricultural landscape is necessary before we consider the good matches occurring at Scenic View.

As the previous chapters have outlined, there have been broad changes in, and challenges to, the organic movement throughout its history, from its early anti-industrial roots to its modern incorporation into the pantheon of products produced by agro-industries. Scenic View participates in that story. As a small farm that chooses not to sell to conventional retail outlets (as we will see in the chapters that follow), Scenic View is not as embedded in the industrial organic food chain as other farms of even similar size. Understanding how a small farm like Scenic View maintains its alternative practices requires an understanding of not only the structural openings a bifurcated organic sector provides for farms like this (see Rosin and Campbell 2009). It also requires an understanding of how farmers make choices about which agricultural practices they utilize, how they manage their farms and landscapes, and what marketing opportunities they pursue in order to both stay in business and reflect what it means to them to be organic farmers.

The challenges to staying in business and making a living that today's farmers face are reflected not only in farm loss, but also in the demographic composition of the farmers that remain. Today, "American farmers over the age of sixty-five outnumber those under thirty-five by nearly six to one" (McKibben 2007, 55). According to the USDA's 2007 census, the average farm operator in the United States was fifty-seven years old. In Rhode Island, Massachusetts, Vermont, New Hampshire, and Maine the average age was slightly younger, about fifty-four. In Connecticut the average farm operator was about fifty-five (USDA 2009). Today, the average age of America's farmers continues to rise. According to the most recent census in 2012, the average principal farm operator in the United States is over fifty-eight years old (USDA 2014). While the number of farming operations has declined and agricultural production has concentrated and intensified, the farmers remaining in our food system have faced a graying future.

Scenic View Farm, however, represents part of an emerging counter-tendency within America's agricultural system. When we encountered Scenic View Farm, the 2007 Census of Agriculture showed, for the first time since World War II, that the number of farms in the United States had increased. Between 2002 and 2007, there was a 4 percent increase in the number of farms, amounting to about 76,000 new farming operations. Moreover, while the historic trend has been toward fewer, larger farms, the majority of those new farms were on the smaller side, with young operators who also worked off the farm to make ends meet. During that period, the number of farms with sales under $1,000 per year increased by 118,000 (USDA 2009).

Today, we know that the uptick in the number of farms in the United States was short-lived, perhaps nothing more than a blip in the historic downward trajectory. In 2012, there were approximately 3 percent fewer farm operators in the United States than in 2007, representing the loss of 101,460 farmers. The loss was greatest for primary farm operators, with a drop of about 4 percent. There were also fewer new farmers in 2012, with a 20 percent decline in the number of farmers who had been on their farms for less than ten years. However, one category of farmer continued to increase through 2012: young, new farmers who consider farming their primary source of employment. Between 2007 and 2012 there was an 11 percent increase, or just over 4,000 young people, getting into full-time farming—farmers very much like those we encountered at Scenic View during the 2009 growing season (USDA 2014).

While the number of small farms and new farms has been in flux, concentration and fragmentation have remained consistent, defining features of the contemporary American food system. In 2007, 125,000 farms—or approximately 5.7 percent of American farms—produced 75 percent of the value of U.S. agricultural production. In 2002, 144,000 farms produced the same value of production (USDA 2009). Today, 75 percent of American farms have annual sales under $50,000, and the total agricultural output of these farms contributes to merely 3 percent of all farm sales (USD 2014). Moreover, in 2012, 50 percent of all harvested acres were devoted to just two agricultural commodities: corn and soybeans (ibid.). Even if young people are continuing to try to break into agriculture as a full-time occupation, they are still coming up against an agricultural sector dominated by bigger and bigger farms specializing in fewer and fewer crops. Even outside of the organic market, small farms in general exist in an increasingly bifurcated food system; fewer, larger farms continue to capture more of the market, while smaller farms face a far less certain trajectory.

Despite such persistent trends, these same data show the ways many contemporary farmers are trying to build new types of economic relations on their farms outside industrial food supply chains. Scenic View Farm is part of a growing number of farms choosing to market directly to consumers. In the period 2002–2007, the growth in direct sales of farm products far outpaced the total growth of agricultural sales. According to the 2007 agricultural census, direct sales increased 104.7 percent, while total growth was only 47.6 percent (Agricultural Marketing Service 2009, 3). Today, direct-to-consumer sales total approximately $1.3 billion—up 8 percent since 2007 (USDA 2014).

It is interesting to note that direct sales are particularly strong in New England, with all six New England states appearing in the top ten states in terms of direct sales (Agricultural Marketing Service 2009). As we have seen, New England has a history of market gardening (Bell 1989), and, more recently, the region "has [supported] the heaviest concentration in the country of retail farming operations; moreover, at least 75 percent of the population is within ten miles of a retail outlet selling fresh produce on a seasonal basis" (Jager 2004, 53). As Brian Donahue (1999, 90) has suggested in New England's agricultural context, market-garden farming operations "can hardly help paying" once the exorbitant cost of the region's land is paid off—especially when they can offer more connected consumer experiences.

Direct marketing can take several forms, although the most prevalent is the farmers' market. The growth of farmers' markets has been explosive. There were over 2,800 farmers' markets in the United States in 2000. In 2009 there were more than 5,000, and by 2011 the number had risen to over 7,000 (Agricultural Marketing Service 2011). By 2014, the number continued to rise to over 8,000 (Economic Research Service 2014). Similar trends have occurred in the growth of CSA programs. In such programs, customers do not merely buy vegetables. Rather, like the stock market, they buy shares. However, CSA consumers purchase shares of a farm's potential produce for a season. Moreover, on a farm like Scenic View, operating a CSA is perhaps one of the most important ways small farmers seek to build alternative social relations around food and make a life in agriculture consonant with their personal values. As the subsequent chapters will show, many of the most potentially transformative relationships at Scenic View were the results of matches afforded by the CSA model. More important, many of the struggles Scenic View and other farms like it continue to face in their efforts toward producing greater agricultural sustainability could be addressed by strengthening and expanding the types of relationships CSA programs offer, making them a critical tool for building a more organic future and a more environmentally conscious economy.

According to the USDA, the CSA concept originated in Switzerland and Japan in the 1960s, "where consumers interested in safe food and farmers seeking stable markets for their crops joined together in economic partnerships" (DeMuth 1993). The first CSA in the United States was started in Massachusetts in 1985 (McKibben 2007, 81). By 1993, the USDA estimated there were over four hundred CSA programs in the United States (DeMuth 1993). Although the USDA does not regularly keep track of the number of farms utilizing this marketing strategy, Local Harvest (2015), an online directory, lists over six thousand farms offering CSA subscriptions. In just over twenty years, the number of farms marketing with CSA programs has risen well over tenfold. At the same time, studies of bifurcation in the organic sector have shown that "movement-oriented" farms tend to market directly to consumers more than "minimalist," input-focused farms (e.g., Constance, Choi, and Lyke-Ho-Gland 2008). Moreover, with the increasing conventionalization of the organic sector, consumer demand is increasingly shifting toward "local" foods and direct markets. In fact, local has surpassed organic as the fastest-growing food sector (Adams and Salois 2010). Direct-marketing initiatives provide structural space for more holistically organic farms, allowing for a niche market where agro-industrial firms will have relative difficulty competing.

Such trends in direct marketing mark a radical transformation of the American agricultural landscape. Moreover, they present a possibility for small-farm survival: "Locally grown, organic foods help make it possible for today's smaller, highly diversified operations to gross up to $25,000 an acre, whereas in the past, a farmer was happy to make $2,000 in profit per acre" (Jonsson 2006). Despite the fact that the number of farms continues to decline and American agriculture continues to be characterized by intense economic concentration, it is likely no coincidence that young first-time farmers are continuing to set out to make farming a full-time career. Through direct marketing, small farmers finally have a chance to make it economically. And as we will demonstrate with examples from Scenic View and similar farms, farmers can make decisions about how they approach direct marketing in ways that harmonize good business practices with their own ethical concerns and lifestyle aspirations.

Although potentially lucrative, the decision to adopt a direct-marketing approach requires that farmers make a significant investment of time and energy. "It is widely believed—within the farming community at large—that accessing added value from local food requires an increase in farm labor," something busy farmers increasingly juggling off-farm commitments may not be able to manage (Mount 2012, 108). The decision for a small farm to market directly to consumers requires matching lifestyle concerns with business practices—something that the more economic accounts of the organic sector have been critiqued for neglecting (Rosin and Campbell 2009). Theories of bifurcation explain why a structural niche for direct-marketing operations is possible in the current organic market; a language of matches can begin to explain how small farmers fill those niches by making lifestyle decisions such as these.

Relational practices and skills are paramount for exploiting these opportunities since direct marketing demands an entirely different set of skills than those utilized by farmers engaging in wholesale operations (Oberholtzer 2009). Farmers who utilize direct-marketing approaches must have strong interpersonal skills to forge relationships with customers, and they must possess significant knowledge concerning the cultivation of a wide variety of crops to satisfy changing consumer demands. These skills are vastly different from those required to run a successful monoculture operation and shrewdly negotiate a wholesale price with a distributor.

Even with time and expertise, not every farm will be able to successfully market its produce directly to consumers because farms need a large base of potential consumers in order to engage in direct marketing, a factor that is lacking in much of rural America. Farmers in rural areas "have more difficulties reaching consumers, including problems accessing urban markets, and transportation issues" (Oberholtzer 2009). While farms located on the edge of urban areas are able to tap into direct markets, they often face tremendous challenges with the expense of owning or renting farmland on "the urban fringe." Despite the remarkable growth of direct marketing, there are serious obstacles facing farms attempting to access these markets across the United States.

The agricultural economy of New England was once predicated on smaller farms on the urban fringe (Bell 1989; Russell 1976). Even today, the region is well suited for them—as well as the opportunities for direct marketing they allow. As Ronald Jager reports, if measures of agricultural productivity were redefined to account for New England's short growing season, the region would be at the forefront. Moreover, "because the New England landscape is "a hodgepodge of small communities interspersed among pockets of tillable and fertile soil, diversified crop production and marketing" can remain "very closely and intricately intertwined" (2004, 54). For farms that can manage to gain access to direct markets and make the types of good matches that will enable them to viably balance the time and expertise required to succeed, these emerging markets present opportunities.

Such trends have allowed individuals with farms on the "urban fringe"—like Scenic View Farm—to pursue a career in agriculture. However, there is a caveat to such seemingly promising trends. Even when there was an uptick in the number of farms in 2007, most of the growth the USDA census revealed was in small farms with sales under $1,000. According to Mike Duffy, an agricultural economist in Iowa, these small farms are most often "lifestyle" farms, where "people are not trying to be a commercial farmer [sic]" (cited in Jonsson 2006). In fact, the number of farms selling between \$1,000 and \$249,999 also continued to decline between 2002 and 2007, consonant with the declines agriculture continues to experience today (USDA 2009). As a result, we must be cautious when arguing that "small farms" are making a comeback in contemporary American agriculture.

At the very least, we know that farms like Scenic View are not alone. Rather, they are part of a vibrant group of similar small, diversified farming operations utilizing direct marketing across New England and the United States. But however vibrant these farms may be, we also know that the contemporary agricultural sector has faced continued decline, fragmentation, and consolidation. By paying attention to the ways in which the farmers at Scenic View negotiate the challenges and opportunities of today's agricultural economy, we can gain important insight into what these broader trends mean for those experiencing them. The unique ecological constraints of farming that undermine the profitability of industrial-scale farms in regions like New England, the relative productivity of small farms in some circumstances, and consumer concerns about the safety and quality of industrial foods provide opportunities for small farmers to craft practices and lifestyles that resist the conventional model of agriculture that continues to dominate the contemporary agricultural economy (Coombes and Campbell 1998, 141; see also Constance, Choi, and Lyke-Ho-Gland 2008). What remains to be explored is how farmers make the types of good matches that allow them to make use of these structural niches.

Scenic View is one farm among many in the contemporary bifurcated organic market pursuing a more alternative vision of organic agriculture than that of supermarket organic. We cannot conclude that every small-scale organic farmer across the nation, or even across the region, shares the same challenges to his or her farm's viability, but we can be confident that the language of good matches we develop through our case study is transferable and fills an important gap in our understanding of bifurcation. In a bifurcated organic market, small-scale farmers must make a match to engage both in farming practices that are more holistically organic than the input-substitution approach of conventionalized farms and in business practices that are profitable. Other farmers may respond to these and other challenges and opportunities in different ways. However, Scenic View provides an illustrative case for farmers who must negotiate alternative social and economic relations for organic. Good matches provide a realistic representation of the ways small farmers struggle to achieve a viable balance among their expectations for their businesses in the face of the diverse forces constraining small-scale organic farmers across the United States.

As noted, this book will also bring to the fore the experiences of other farming operations both in New England and in other parts of the United States. Through our interviews in organic, low-spray/ no-spray, and conventional farms in New England, we will trace—as C. Wright Mills argued all good research should—"history and biography and the relations between the two within society" (2000, 6). We do not include these to describe the New England organic sector in a generalizable way. Instead, we offer these additional accounts to demonstrate the critical importance of social relationships in generating the different types of matches that farmers are making across New England and the United States. In making such matches, these farmers not only occupy but also actively build alternative spaces within the organic sector, as they struggle to both compete with and stay out of the industrial food system. However, "Neither the life of the individual nor the history of a society can be understood without understanding both" (3). The same is true for organic farming. What remains, then, is an analysis of how these forces are perceived, responded to, and matched to specific everyday practices by individual farmers. To understand their daily reality, one has to spend some time in the field.

four

**AMID THE CHARD: CULTIVATING
THE DIVERSE LANDSCAPES AND PRACTICES
OF A NEW ENGLAND ORGANIC FARM**

It is useful to begin this chapter with a brief overview of Scenic View Farm. From there, we can begin to unpack the data and describe the case study site in an ethnographic style so that the relational work, everyday practices, and matches made by the farmers to achieve a livelihood while holding to sustainable ideals can be looked at more closely. Scenic View Farm is a USDA-certified organic farm situated on the "urban fringe" about sixty miles outside of Boston. It is a "family farm," which its principal operators, John and Katie, rent from John's parents and run together. While Scenic View has been in John's family since the turn of the last century, it had not been used as a farm for many years. However, the land has been kept affordable thanks to a conservation easement John's grandmother entered into, allowing some agricultural use of the property but preventing further commercial or residential development of the land. Scenic View is smaller than the average farm in the region, with four fields totaling just over 6 acres under cultivation. In contrast, the 2012 Census of Agriculture found that the average farm size in Massachusetts was 68 acres, while in Vermont it was 171 acres. The average farm in Rhode Island was smaller, averaging 56 acres (USDA 2014). Scenic View's net sales are higher than those of most farms in the region, averaging approximately $85,000 annually. In contrast, 75 percent of U.S. farms have sales under $50,000, and many of those farms are concentrated in New England and the Southeast (USDA 2014). Like most of the organic farms in New England, Scenic View is engaged in crop, rather than livestock, production. In Massachusetts, for example, 264 of the state's 319 organic farms were involved in crop production in 2007 (USDA 2009). Scenic View grows a large variety of vegetable crops, as well as herbs and cut flowers. John and Katie market most of their produce through the farm's CSA operation. The rest of its produce is sold directly either to consumers through its farm stand or to restaurants. In the 2007 Census of Agriculture, only 3 percent of farms in Massachusetts reported having CSAs (ibid.). As we discussed in the prelude above, New England has the most robust direct marketing of agricultural products of any U.S. region. The 2012 Census of Agriculture revealed that CSA programs had expanded to 6 percent of the farms in Massachusetts, making it the state with the highest percentage of farms utilizing CSA arrangements to market farm goods (Keough 2014). As we will see, Scenic View's CSA program has been vital to its success and its ability to make good matches.

Also like many New England farms, Scenic View relies on non-agricultural income to supplement its farm sales. In Massachusetts, for example, approximately 53 percent of farmers did not consider farming their primary occupation in 2012, and 40 percent of farm operators worked two hundred days or more off-farm. The same was true for Connecticut: nearly 57 percent of farm operators did not consider farming their primary occupation, and about 43 percent worked off the farm two hundred days or more in 2012 (USDA 2014). At Scenic View, farmworkers are seasonal, but the principal operators work full time on the farm. However, they do generate non-farm income through a rental business. The property has a cottage and guest house that are rented out for weekend getaways and various other functions. This type of business is possible because of the farm's natural surroundings, making it a beautiful place to hike, explore, and just relax. As we will see, this type of supplemental income allows the operators of Scenic View to maintain the type of involvement with the farm they desire while generating a necessary source of additional revenue.

These characteristics will be revisited, as they arise in the ethnographic account of Scenic View Farm that follows. While these data situate Scenic View Farm in the context of New England farming, they do not capture the pragmatic concerns, sentimental attachments, beliefs, or values that shape the everyday agricultural practices found therein. For these, it is necessary to delve into the practices, stories,

and life experiences hidden in such a thumbnail sketch of a farm
like Scenic View. Moreover, it is necessary to understand and observe
how the people who do the work of organic farming make sense of
the often contradictory pressures they face in their lives. We need to
understand how the farmers themselves make good, organic matches.

As owners, John and Katie are the individuals who bear the primary responsibility of cultivating the fields of Scenic View Farm.
Both are New England natives, both are college educated, and—
despite different life histories—both have become devoted to full-
time careers in agriculture. While John moved back to his family's
land to start the farm, Katie married into a life of agriculture. However, neither John nor Katie were farmers before moving to Scenic
View. John has a liberal arts background, and Katie was once a professional dancer. Despite lacking active farming backgrounds, at
Scenic View Farm they have built a farming business that they had
been successfully growing over the ten years prior to our meeting.

Katie and John's farm sits on a large piece of property; however,
the six acres they cultivate seem even smaller if we compare their
farm to the modern standards of agriculture rather than the average
farm size in the region. As noted in chapter 1, in 1972 Earl Butz
proclaimed that farms should "get big or get out," and this proclama-
tion remained the predominant view of U.S. agriculture for decades
(Duscha 1972). Under the administration of George W. Bush, U.S.
Undersecretary of Agriculture for Rural Development Thomas Dorr
suggested, "The right scale for farms in the future will be about
200,000 acres of cropland under a single manager" (cited in McKibben 2007, 56). Such thinking is not trivial. It provides the context in
which John's family got out of agriculture to begin with and the context John attempted to break into when he began his career farming
at Scenic View.

Then again, scale is all a matter of perspective. John's farm used
to have two fields under cultivation; now it has four. With those fields,
and the possibility of adding another field in the future, John sees
the farm as "small, on the verge of becoming medium." In light of

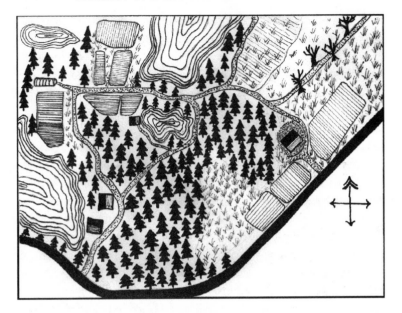

Figure 3. Scenic View Farm: This sketch, drawn from Fitzmaurice's field notes, depicts the agricultural landscape of Scenic View Farm.

Dorr's suggestion of proper farm size, a farm like Scenic View is dismissively miniscule.

Scenic View, however, also is beautiful in its diversity. Anyone who has ever driven through the great agricultural states can understand what a two-hundred-thousand-acre farm looks like. While the seas of corn in Iowa and the majestic wheat fields of the Palouse in eastern Washington possess the power to captivate the eye, Scenic View is more like a nature preserve than a blanket landscape. Indeed, the conservation easement protecting the land from development is a wild lands trust. The farm fields were only one component of the landscape, nestled into the place.

The entire farm is composed of interlaced meadows, swamps, woods, streams, ponds, and agricultural fields (see figure 3). So many birdsongs can be heard while one is working that identifying them is a popular way of passing time in the fields. Herons fly overhead, their immense wings casting shadows on the ground, on their way to stalk fish and small frogs in the ponds. Catbirds can be seen watching the comings and goings of the farmworkers, and baby toads the size of mere pebbles hop frantically to avoid the hands and knees of people weeding between the rows. Owls can be heard calling from the dark woods at night, and one can sometimes even glimpse a coyote as it runs on the thickets in the field margins. On a day in early June 2009, one of the farmhands was driving his truck between fields when he stopped to get out. He spotted a large snapping turtle blocking the path. Tractor drivers have to be careful when plowing the fields since the soft, rich soil is a favorite nesting place of these cantankerous reptiles.

The remarkable beauty of the place was one of the factors that drew John to a life of farming. As a kid, John spent time on his grandmother's farm, working at the horse farm during summer vacations. After college, John lived and worked in a large city but had an itch to get back to his family's land. "Who gets to live in a place this beautiful?" John remembers thinking. In his heart, John felt as though he should live and work his family's land; if not, he would end up facing the onerous responsibilities of ownership one day without ever having experienced the many pleasures and benefits of the place. John has been enjoying the pleasures and dealing with the many troubles of farming ever since.

The beauty of the farm—and the outdoor lifestyle it affords—struck a chord in Katie too. Katie's New England childhood often brought her in contact with the natural landscape, and she fondly recalled childhood summers spent in the woods. As a result, a life in agriculture was not totally alien to her when she moved onto the farm after her marriage to John. Nevertheless, transitioning into a lifestyle of farming meant significant changes in Katie's life. The demands of running a farm left little time for her to pursue her previous career as a dancer. However, it also meant she no longer needed to patch together an income with temporary office jobs—a change for which she seemed grateful. Katie remembered hating office work and felt that spending her summers outdoors farming was a better fit.

While the farm is situated in a scenic part of New England, much of its beauty also comes from the way that John and Katie have managed the land. With farm fields nestled seemingly naturally into the landscape, the rows of crops do not create a sense of endless agriculture. Rather, they seem integrated into the landscape, adding variety while complementing the property's natural qualities.

While these characteristics are the result of the farm's management choices, they are not unique to Scenic View. Rather, many studies have shown that organic farms tend to produce more sustainable rural landscapes. "Organic farms offered more variety and variation [than their conventional counterparts], yet were visually coherent on the landscape" (Duram 2005, 56). And they provide both healthful local food and a beautiful landscape, what Donahue calls not "[a] bitter but a delicious offering on our plate. We can offer the better flavor and purity of locally grown organic food and the beauty of farms conveniently located in customers' backyards. . . . We can sell the ambiance of the farm itself" (1999, 75). Ecological studies of biodiversity on farms have shown that the sustainable landscapes of organic farms are more than merely aesthetic. For one thing, as noted, organic farms are more diverse. One study found significantly more butterflies on organic farms (Feber et al. 1997). Another found twice the number of bird species, and more than twice the number of birds, in organic farming systems when compared to the conventional (Beecher et al. 2002).

For many of our interview participants on other farms across New England, it was the beauty of the agricultural landscape—and the possibility of living—or at the very least working—"on a place this beautiful" that sold them on a lifestyle of farming, despite the economic hardships that so often accompanied such a choice. (See table 1 for a list of interviewees and farm demographics.) Consider Elaine, a field hand and farmers' market salesperson for Pioneer Farm, a thirty-acre certified-organic farm with a 250-member CSA that has

Table 1: Interview Sample Demographics

Farm Name*	Size (Cultivated Acres)	Organic Certification Status	Principle Crop(s)	CSA Membership
Scenic View Farm	**6**	**Certified organic**	**Mixed vegetables, cut flowers, herbs**	**100**
Viridian Farm	50	Non-certified, organic methods	Mixed vegetables, sunflowers	750
Old Times Farm	15	Non-certified, organic methods	Mixed vegetables, chickens, sheep, turkeys	22
Peaceful Valley Orchards	130	Integrated Pest Management	Mixed vegetables, orchards (apple/peach)	N/A
True Friends Farm	60	Integrated Pest Management	Mixed vegetables	<100
True Friends Farm	60	Integrated Pest Management	Mixed vegetables	<100
Happy Hen Pastures	80	Non-certified, sustainable/pasture-raised	Laying hens, chickens, turkeys	100
Verdant Acres Orchards	250	Non-organic	Mixed vegetables, orchards	N/A
Heritage Harvest Farm	65	Non-certified, sustainable and organic practices	Mixed vegetables, eggs, honey	510
Pioneer Farm	30	Certified organic	Mixed vegetables, orchards	250
Sustainable Harvest Farm	130	Certified organic	Mixed vegetable	1,500
Long Days Farm	144	Sustainable/Integrated Pest Management	Mixed vegetables, livestock, orchards	300
Johnson Family Farm	35	Low spray/no spray	Mixed vegetables, fruit, firewood, poultry	<200
Parnassus Farm	55	Certified organic	Mixed vegetables	N/A
Grateful Harvest Farm	13 (plus pasture)	Certified organic (Non-certified pasture)	Mixed vegetables, livestock,	100
Laughing Brook Farm	20	Certified organic	Mixed vegetables, berries, apples	N/A

*All names provided are pseudonyms.

Other Sales Venues	Labor	Participant in Our Study	Participant's Position
Restaurants, farm stand	**Apprentices, full- and part-time seasonal workers, one full-time farm/ rental manager**	**All employees: case study site**	**All positions**
Farmers' markets, urban farm store, restaurants	Apprentices, full-time farm managers	Collin	Farm owner/ operator
Farmers' markets, restaurants	One full-time employee, international interns	Sally	Market manager/ field laborer
Farmers' markets, farm stand, restaurants	Seasonal workers, farm managers	Matthew	Market manager/ grower
Primarily wholesale, farmers' markets	Family, seasonal workers	Mary	Farmers' market salesperson (neighbor)
Primarily wholesale, farmers' markets	Family, seasonal workers	Jane	Field intern, market salesperson (daughter)
Farmers' markets, farm stand, local independent grocery stores	5 local women for hand butchering, market manager	Susanna	Market manager
Farmers' markets, farm stand, pick-your-own, wholesale (apples)	Family, seasonal workers	Sean	Field worker/market salesperson
Farmers' markets, farm stand, limited wholesale	15–20 field/market workers	Claire	Former field worker/market salesperson
Farm stand, restaurants, wholesale to local food hub	Seasonal field workers	Elaine	Seasonal field worker/market salesperson
Farmers' markets, farm stand, wholesale to restaurants and coops	Seasonal field crew, market salesperson, 4 interns	Meghan	Seasonal field crew member/market salesperson
Farmers' markets, farm stand	Unspecified	Lucy	Farm owner/ operator
Farmers' markets, farm stand	Few seasonal workers, farmer operator	Eric	Farm owner/ operator
Wholesale, farmers' markets, farm store	Seasonal field workers, market salesperson, wholesale/packing house workers	Rita	Market salesperson
Farmers' markets, restaurants	Unspecified	Molly	Farm manager
Farm stand, farmers' markets	Seasonal field workers, market salesperson	Helen	Market salesperson/ field worker

been operating for the past five years. When asked about why she decided to start working in agriculture, her voice lowered and notice- ably slowed with palpable satisfaction as she remarked, "Oh, it's a beautiful farm." Collin, who owns and operates a farm with a seven- hundred-member CSA, recounted how he "fell in love with the small-farm aesthetic" while interning on farms during college.

Then there is Sally. For the past nine years Sally has been the only full-time field worker and market manager at Old Times Farm, a non-certified farm using organic methods with fifteen acres of cul- tivated land. As a new mother, she began volunteering on the farm for a few hours a week as part of her CSA membership. She fondly remembers that "it was a beautiful spot . . . it was pleasant" to be in the fields with a toddler and that she loved the "atmosphere." It was the atmosphere of the farm itself that made her willing to accept the position when the farmer offered her employment the next season. She has been farming every season since.

The springtime approach by vehicle to Scenic View Farm is like that for a quintessential New England farm. First, one must pass sev- eral houses along a shaded road that still belong to John's extended family, then cross a small stream that flows into a large pond on the left of the property. The first field in view is relatively small, filled with rows and rows of flowers, not yet in bloom, and then a variety of herbs (see figure 4). After one turns a bend, four more fields can be seen with small dirt roadways in between, the fields surrounded by grass and bordered by dense stands of trees. Still early in the season, the brown, sandy soil is marked by the ridges and furrows created by tractor blades. This place was alive with diversity. Stepping onto the land, one can see the peas beginning the slow climb up their trellises, the carrot tops pushing their lacy fronds through the soil, and, in the distance, the dark green leaves of potato plants. To the right one sees row upon row of beautiful young heads of lettuce, Swiss chard, beets, and beans. However, the life on this farm was not limited to just vegetables. Bird calls sounded across the fields, coming from

Figure 4. New England Pastoral: The landscape of organic agriculture at Sustainable Harvest Farm. (Photo courtesy of Connor Fitzmaurice.)

the tress. Sparrows chirped cheerful tunes while catbirds screeched out their harsher notes.

Amid the diversity of flora and fauna one finds people working to bring forth new life at Scenic View, cultivating tiny seedlings shooting up from the cool soil characteristic of New England springs, and reflecting ideals central to the pioneering conceptions of organic farming. Describing organic agriculture, the organic pioneer Lady Balfour remarked, "Though I prefer the term biological husbandry because of its emphasis on life, the short answer is balance" (Balfour 1977). Unlike farming monocultures, the landscape of Scenic View seemed to be an attempt to achieve balance.

Deeply ingrained in the ethos of organic farming is the view of farms not as sites of mere economic resource extraction but as living ecological systems. "For the organic farmer, the key concept is diversity and the health of the soil. For the conventional farmer, the key concept is monoculture and maximum short-term profit" (Blatt 2008, 86). According to agricultural economist John Ikerd, these divergent practices arise from the differing logics underpinning what he terms "permanent" and "productive" agriculture. Permanent agriculture is based on the natural order, endlessly and self-sufficiently recycling wastes and maintaining fertility. Diversity is central to such systems. In productive agriculture, diversity gets in the way. The goal is not permanent maintenance but maximal resource extraction. The results of the former are closer to sustainability; the results of the latter are certain decline (Ikerd, in Kristiansen, Taji, and Reganold 2006). The diverse array of crops sharing each bed seemed to follow nature's contours and conditions, fading into and out of the surrounding fields and woodlands.

The organic method presents all manner of obstacles that the conventional farmer can avoid with certain chemicals. For example, John explained that the incessant drizzle typical of New England springs made cultivating the fields difficult. If weeds are pulled in the rain, they just set down new roots in the moist soil. However, farm work can't wait on the weather, for it is often uncooperative. So one must take a hoe and set to work with the most basic of farming equipment. In the springtime workers drag the narrow, freshly sharpened blades as close to the tender new bean sprouts as possible, without mowing them down, of course, to ensure that the greatest number of weeds are uprooted. Such weed pulling is critical since beans are difficult enough to harvest without a bunch of weeds getting in the way. Harvesting beans, however, was a prospect for the summer.

Spring was the time for preparation of the soil and the nursing of young plants.

Making Hard Choices to Grow Unprofitable Crops

As we worked through the beans and weeds on Scenic View one day with John, it brought forth fascinating insights that explained the hard choices—and good matches—that organic farmers make every season. The idea that economic rationality is separate from, and hostile to, intimacy and sentiment flaunts the evidence of how people experience their economic lives (Zelizer 2010). Instead, people attempt to match the economic and sentimental commitments of their lives in ways that get "the economic work of the relationship done" while sustaining the *character* of the relationship (Zelizer 2006, 307). Organic farmers are in precisely the types of relationships where good matches become critical; the goal of matching is to prevent one type of relationship from being mistaken for another "with which [it] might easily and hurtfully be confused" (315). How, amid the stress and promise of spring's first bean sprouts, do farmers like John and Katie ensure that their need to stay in business does not come across as their being in it for the business?

The bean shoots had only recently emerged from beneath the soil, and their first sets of leaves were just beginning to unfurl. Some, however, were clearly stunted. The stems stood bare, stretching toward the sun in vain. Every few feet, a bean sprout stood leafless. While walking down the rows, disturbing the soil with the hoe blade, John uncovered the presumed culprit. He stooped down and lifted a small white and yellow grub into the air—a cutworm.

John explained that cutworms are a constant plague of vegetable plants. Living beneath the surface, the subterranean creature wraps itself around the tender new stalks as they force their way upward toward the light. Looped around the shoot, these grubs chew, severing the stems at the point where they attach. This activity explained the lack of leaves; the tops of the bean sprouts were being eaten before the plants even had a chance to emerge. This destructive creature called for the use of another age-old device necessary to the organic farmer. Before walking away, John crushed the writhing grub between his fingers.

Continuing down the rows of beans, John expanded on his analysis of the importance of keeping the weeds under control. Not only are beans tedious to pick, but they are also unprofitable. This fact might be rather shocking to some readers. In the grocery store, beans seldom appear particularly cheap—unless of course they come in a can. Fresh organic beans always cost at least a couple of dollars a pound, and the delicate younger beans cost even more. "Well, that's sort of the problem with having a CSA," John said. "You have to grow cheap things because that's what people want. If I could [grow one crop], I would grow only tomatoes. Each one weighs about a pound, and you can charge three dollars a pound. Even if you could get the same price for beans, what does each one weigh? One one-hundredth of a pound? It's a lot more work picking them than it's worth." With a chuckle he concluded, "But I don't really see a tomato-only CSA being all that successful."

Between the maroon-streaked beet greens and the slender bean sprouts that seemed so natural together in John's fields, there was an underlying economic as well as ecological logic. The diversity of crops was neither purely accidental nor entirely ideological; it was a good match that both suited the organic farming principles to which the farmers at Scenic View were committed and allowed them to engage in a business model that helped their farm stay afloat.

Business practices such as these demonstrate the careful negotiation of the economic challenges of running an organic farming business with the concerns that bring farmers into organic agriculture in the first place. It is immediately clear that Scenic View fits the "lifestyle-farming" niche of bifurcation approaches (see Constance, Choi, and Lyke-Ho-Gland 2008). It is also clear that this fit is the result of purposive negotiations about what type of farm its owners hope to maintain. As a result of John and Katie's conscious decisions, for example, Scenic View has eschewed any wholesale ventures that would put their products on retail store shelves. Instead, they rely upon their CSA as their main form of sales—accounting for 60 percent of their business, or about $48,000 in net sales—a decision with everyday repercussions for what they grow on their farm, as well as how and when they grow it. John and Katie are not alone in their decision to avoid or limit their participation in conventional market channels. Rather, we can see clear parallels in the marketing practices of the other farmers with whom we spoke across New England. Increasingly, CSAs have become one of the few ways small farms can make ends meet, by guaranteeing purchases at prices capable of sustaining the farms through the season. A "tomato CSA" might have enticed John, but he knew it was more than merely ecologically unsustainable.

Other farmers echoed John's sentiments when they described the hard decisions they made to grow often unprofitable crops. Collin, the owner of Viridian Farms, recounted a story similar to John's. After apprenticing on several farms during and shortly after graduating from college, Collin began his own small farm on ten acres of his parents' land. After considerable struggle to stay afloat, he was forced to close his fledgling farming operation after four seasons. Eight years ago he began farming again, and his farm has been growing by five to ten acres a year and boasting a seven-hundred-member CSA, an urban farm store, and significant farmers' market sales. Despite the flourishing CSA membership, Collin explained the curious fact that it is sunflower sales at farmers' markets that are the "cream crop" of his business.

The CSA, Collin explained, is important for providing up-front cash and redistributing agricultural risks to consumers. It also gives the farm flexibility throughout the season since customers are provided a weekly "mystery basket" rather than being promised precise quantities of particular crops. If the pepper crop performs poorly during the season, for example, other crops can be substituted for peppers. However, despite the benefits of the CSA model and Collin's twenty years of farming experience, he chuckled a bit when he explained, "We always figure that the sunflowers are the *one* thing we actually make a profit on. Everything else just covers costs. We are still trying to figure out how to make the vegetable production profitable too."

Increasingly, farmers are finding that it is not economically feasible to rely on monocultures and that diversity is essential, even if diversity means growing less profitable crops in exchange for the security a CSA model can provide. Lawrence Mendies of Roosevelt, New Jersey, learned such a lesson the hard way. Unable to sell an abundance of zucchini one summer, Mendies brought his bushels to the local auction. He explains: "The highest bid was $2.50 per bushel. I had so many on the truck. I did not know what to do. The baskets the zucchini were in cost $1 each. I got $1.50 each. The time it took to grow and pick each bushel . . . I had to sell them" (cited in Kohlhepp 2011, 27). Mendies now operates a successful CSA, much like John and Katie at Scenic View Farm.

So although some crops grown on farms like Scenic View and Viridian are unprofitable, they are necessary to give CSA customers the diversity they expect in their baskets. Other crops, however, serve as useful fillers during the less productive months of the cool New England spring. Scenic View grows chard as such a springtime filler for its CSA boxes. The chard plants look majestic against the backdrop of a rainy spring day. After months of winter, the stalks shoot forth from the rich soil in multiple colors: pink, orange, yellow, green, white, and crimson. Chard grows wonderfully in the cool damp New England springs. And in June 2009 it rained nearly continuously for the entire month. When it hadn't been raining, it had been unseasonably cool and damp—again, perfect weather for chard. Unfortunately, the ideal weather for chard was a problem for some other, more profitable crops.

Tomato Blight and the Plight of the Organic Farmer

The rows of beans and Swiss chard reveal one of the complex negotiations organic farmers undertake when making seemingly trivial decisions—like determining how many varieties of beans to grow or whether to grow any at all. The rows of tomatoes reveal another.

"You may have picked an interesting time to write about our farm," John commented wryly one late afternoon in July. "It could be good for your research but bad for us. With all of this rain, the chances of disease are much higher, and since we're organic, there isn't all that much we can do. It could be an interesting season."

John went on to explain the options, the possible matches he and Katie could make. If the farm were conventional, they would have been able to spray fungicides to cure the plant diseases that were likely to spring up in the unseasonably damp weather. Organic farms could use a copper sulfate solution since it was exempted on the National List of Allowed and Prohibited Substances; however, John viewed the use of copper sulfate as an unsustainable method. As we indicated in chapter 2 above, the exemption for copper sulfate is a potentially problematic allowance; copper sulfate accumulates in the soil over time, leading to levels of toxicity that John, Katie, and many organic farmers and observers think ought to be avoided. As a result, that left Katie and John with the creative method of spraying beneficial bacteria on the plants; these bacteria populate the leaves, possibly competing with the harmful fungi for space. However, while creative, this method is not as effective a suppressant as fungicides would be. Rather, if the conditions are very poor, it is only a means of forestalling the inevitable. However, matching the possible economic loss of the tomato crop with their skepticism about the safety of copper sulfate—despite the fact that it is technically organic—John and Katie viewed forestalling the inevitable as a workable solution.

By early July, the farming community and the public at large were abuzz with talk of tomato blight. It seemed as though John's prophecy was coming true: with the cold, damp weather in June, farms across the Northeast were in for a challenging season. The news even made it into the *New York Times:* late blight was prematurely striking tomato crops throughout the region.

One afternoon, a woman came to Scenic View to pick up some vegetables and pointed out the blight on the leaves of some of the tomato plants. Everyone wondered what the blight could mean for Scenic View Farm. According to John and Katie, it could mean the loss of a quarter of their income if the tomato crop failed. "A highly contagious fungus that destroys tomato plants has quickly spread to nearly every state in the Northeast and the mid-Atlantic, and the weather over the next week may determine whether the outbreak abates or whether tomato crops are ruined," the *New York Times* would later report (Moskin 2009, 1). Plant scientists now recognize the blight that struck during the cool, damp weather of the summer of 2009 as a new genotype: *Phytophthora infestans*, US-22. The University of Massachusetts at Amherst's Agricultural Extension Service information page on late blight reported that in 2009 "growers were required to make numerous fungicide applications of the best materials on a shortened schedule to harvest marketable fruit" (UMass Extension 2012). This statement, however, is not the whole story. Conventional farmers in the region were able to make numerous fungicide applications to harvest marketable fruit. But as John and Katie had worried in the days before the epidemic made headlines, such applications weren't an option for them or the countless other organic farmers in the thirteen states affected (O'Neill 2009).

Stories abound from that summer among farmers, revealing the environmental and economic tightrope small organic farmers must walk—and the often fraught conditions under which they seek to make good matches. For example, Barber noted that one Hudson Valley farm lost half of its field of tomatoes in just three days. The rapid advance of the disease left little time to plan for the worst and harvest crops early. "Organic farmers were forced to make a brutal choice: spray their tomato plants with fungicides, and lose organic certification, or watch the crop disappear" (2009). The impact on organic farms was so dramatic that the Connecticut chapter of Slow Food, the international sustainable food organization, was forced to cancel its heirloom tomato festival; there simply weren't enough heirloom tomatoes to go around (Venkataraman 2009).

For some organic farms, the losses were catastrophic. Lindentree Farm, in Massachusetts, lost its entire tomato crop—a half-acre field that characteristically yielded 2,400 pounds of tomatoes a week (Venkataraman 2009). If we assume that this farm sold its tomatoes for roughly the same price as does Scenic View Farm—$3.99 per pound—that amounts to a loss of $9,500 a week or just under $40,000 a month. That is more than many small farms make in an entire growing season. Many farmers, like Amy Hepworth of Milton, New York, were forced to proactively destroy their plants to prevent the spread of the disease—certainly a heartrending loss of income and effort. Hepworth, a seventh-generation farmer, was growing twenty acres of tomatoes in 2009. She was forced to burn her infected plants, hoping to spare those that remained from contracting the disease (Moskin 2009). Among those farmers who made the difficult decision to spray prohibited fungicides and lose their organic certification—Jay and Polly Armour of Gardiner, New York, who had been organic farmers for twenty years—things weren't much easier. They reported that on their farm, "The fruit [was] rotting under the spray" (ibid.). Regardless of how these farms and farmers responded—whether they remained certified organic or not, whether they destroyed their crops or watched and waited—their choices demonstrate the need to take good matches seriously. Just like those at Scenic View, these farmers were placed in a position where their practices were shaped by their attempts at attaining a viable balance between their economic and personal lives.

For John and Katie at Scenic View, the devastation wasn't quite so complete. Perhaps it was a good thing that John never started the tomato-only CSA he mused could be so profitable. The tomato blight resulted in annual sales that were down by approximately $6,000 when compared to the years before and the year after. However, even with the spread of disease, plenty of other crops thrived. Moreover, since CSA members buy shares of potential crops, they also buy shares of potential crop failures. As a result, John's customers helped him to bear the loss, and the sales that suffered were only 40 percent of the farm's business outside of the CSA. Nevertheless, this dramatic incident serves as an example of the struggles small organic farmers face. And while Scenic View Farm made it through, the epidemic certainly left its mark on the farm. Two years later, when we checked in with Katie and John, they were still talking about the blight.

For some of the New England farmers in our study, it was the fear of the outbreak of a disease—like late tomato blight—that could threaten their entire crop that led them to eschew organic certification. For these farmers, it wasn't worth putting money into certification and effort into the required paperwork only to have to make the decision to give up the certification in order to spray and save their crops—like the Armours had to do during the 2009 late blight epidemic. At a farmers' market one afternoon, a customer asked Eric of Johnson Family Farm, located in the suburbs of Boston, about his pesticide use since the farm isn't certified organic. When he said that he sprayed, the customer seemed dissatisfied and appeared prepared to leave empty-handed. But, Eric went on, "I pretty much only spray the potatoes and the eggplants for potato beetles and the tomatoes for blight. Otherwise, I don't spray. I'm not concerned about cosmetic damage. Look at the leaves on the greens. They are full of bug holes. You can tell I'm not spraying those. It's just sometimes you need to spray to not lose an entire crop."

In a similar manner, Matthew, the market manager and grower for Peaceful Valley Orchards with forty-three years of agricultural experience, explained his farm's decision to use an Integrated Pest Management system (where the decision to spray organic and synthetic chemicals is made based on thresholds that assess whether particular pests or diseases threaten an entire crop). He explained that it wasn't cost effective for his farm, with its abundance of orchards, to risk losing crops by not spraying at all. However, "We, like all consumers, want to avoid pesticides and keep them out of our food." To that end, entomologists and plant pathologists come to the farm on a weekly basis to determine whether spraying is "really necessary." As Matthew explained, "Pesticide use is all of our concern, and it's my concern as both a farmer and a consumer. But I think its important to be able to look at people and say I think our balance [of pesticide use] is right and we're being responsible. I mean, I can spray more, and there won't be corn borers in my corn, or I can keep doing what I think is right, and customers can cut the worms out."

Both Eric and Matthew avoided pesticide use whenever possible. They simply thought it impractical to risk having to either lose costly organic certification by spraying or lose an entire crop by adhering to organic regulations when situations like the tomato blight of 2009 arise. While not certified organic, much like Katie and John these farmers are making good matches that allow them to get the work of farming done—because they see the matches both as profitable choices and as choices they can feel good about. Moreover, despite their lack of certification, these farmers are making the type of good matches that could lead to a more sustainable, organic future. Thinking about good matches helps extend our understanding beyond bifurcation by allowing us to see that contemporary organic farmers are not either "movement-" or "market-" oriented (cf. Buck, Getz, and Guthman 1997); good matches help us understand how farmers are often situated between the two.

On Scenic View, the tomato blight reveals that John and Katie's ethical commitment to organic certification—which prevented them from using chemical solutions to even the most virulent problems on the farm—is central to the farm's identity. But it also reveals that their decisions were necessarily motivated by the economic realities of everyday life. In their discussions about how to handle the blight, John and Katie mentioned that they dealt with blight every year and that their losses were never catastrophic; in 2009 the blight just came earlier. They chose to respond in the way they did in equal part because they didn't think it would be the death knell of their business, given their reliance on a CSA model for most of their sales, and because they didn't want to relinquish their commitment to ecologically sustainable practices.

Certainly other farms and farmers must make similar kinds of matches when attempting to hold on to ecological values amid the need to remain economically viable. Brian Donahue explains it well when discussing this very issue at Land's Sake: "To us, *ecology* always comes first: any use we make of the land must be ecologically sustainable in the broadest and hence strictest sense of the term. . . . At

Land's Sake the *economic* imperative is strong but always constrained by the ecological—that is, we cannot allow ourselves to operate in the most profitable way if doing so degrades the long-term health of the land" (1999, 80).

Even farmers in our study who choose to spray limited pesticides on their crops seem to share similar concerns when deciding on practices. The outcomes of the good matches may be different, but the motivation to choose farming practices that are both economically viable and consistent with the types of relationships they hope to cultivate remain the same. As Matthew of Peaceful Valley Orchards put it, "We want to be able to say we are doing the right thing."

In the year of the devastating outbreak of late tomato blight, the ways John and Katie navigated the USDA allowance for the use of copper sulfate in organic agriculture—and their ultimate decision not to use it on their fields—reveal the power of alternative social and ecological rationalities in organic farming practices. That organic standards allow for the substitution of copper sulfate for conventional fungicides on organic farms does not mean that all farmers will make use of that exception. When farmers see such exemptions as being at odds with the commitments they maintain with their consumers, their land, and to a particular lifestyle, these other considerations often take precedence over technical definitions of organic practice. Likewise, even when organic farmers seemingly engage in chemical input substitution, it would be wrong to assume that such activity automatically denotes conventionalization. Given the complex history and processes that have historically distinguished organic farming, the meaning of "organic" is necessarily a spectrum of beliefs and practices eluding easy categorization.

By the first day of July 2009, the awful weather had cleared, and John, Katie, and their two-year-old son sat under the shade of a tree, escaping the noon sun and eating bagged lunches. John was telling us about how he had ended up on the farm and was lamenting that he hadn't known earlier in his life what he wanted to do. Katie protested:

"The other things you did were great experiences. Besides, didn't your parents discourage your agricultural interests?"

"Yeah, but with good reason. No small farms had done anything but fail for thirty years," John replied, leaning back in his chair and lifting his wide-brimmed hat from over his eyes.

There certainly was something to that argument. According to Michael Pollan, whose books on food and agriculture were read by almost everyone on the farm, the decline of the family farm accelerated in the 1970s as the United States entered into the modern era of agribusiness. As we mentioned in chapter 1 above, under the Nixon administration's secretary of agriculture, Earl Butz, farmers were advised to increase their farm size or cease to farm. With such advice coming straight from the government's mouth, how could responsible parents possibly steer their children into agriculture in the vain hopes that they would survive on the family's small plot of land? The world was moving toward farms of two-hundred-thousand-plus acres, after all.

When John had come back to the farm ten years ago, his family thought it was some kind of joke. "In the beginning, it really was a joke," John admits. "I couldn't grow anything," he says with a grin, lifting a forkful of rice mixed with fresh-picked beet greens to his mouth. The irony of time was evident as John ate the fruit of ten years' worth of labor. In the adjacent field, a couple dozen rows of tomato plants were beginning to flourish in the sunlight. The wet weather had done a number on them, but even a couple of sunny days helped them perk up a little. Peppers and eggplants were also beginning to grow vigorously in the warming weather. Beyond the tomatoes, three varieties of summer squash and four types of cucumbers were being nurtured under fine mesh tents to keep the cucumber beetles at bay. Just beyond the barn, another field boasted a variety of crops. More unusual vegetables joined tried and true turnips, radishes, broccoli, kale, onions, and garlic. Edible flowers and an array of Asian greens with names like senposai, yukina savoy, totsoi, and komatsuna flourished there under John and Katie's care.

In the lower fields, separated from these by woods and swampy lowland, more crops were thriving. Carrots, beets, peas, beans, Swiss chard, numerous varieties of lettuce, winter squash, and potatoes filled the fields, while flower beds added further diversity to the farm's products. On these few acres, more varieties of vegetables were growing than fill the average supermarket. In a sense, this isn't all that surprising. The number of edible species—food, in other words—that make up the modern agricultural pantheon is startlingly low: "About twenty pampered plant species make up the bulk of modern agricultural production; eight are grass species" (Blatt 2008, 83).

Such a decline in varieties under cultivation has been in motion for decades, with the National Academy of Sciences reporting in the mid-1970s that "most major crops in the United States are 'impressively uniform genetically and impressively vulnerable'" (1975, 4). This decline, however, has accelerated. As a result, the U.N. Food and Agriculture Organization (2010) estimates that 75 percent of crop biodiversity was lost between 1990 and 2000. In the United States alone, "ninety-seven percent of the crop varieties once listed by the USDA have been lost in the past eighty years" (Blatt 2008, 84). This figure includes 93 percent of the lettuce varieties, 98 percent of the asparagus, and more that 95 percent of the tomato varieties once available in the country (ibid.). Foster (1999, 94) laments that today a mere fifteen species provide close to an astonishing 90 percent of all human energy, and a mere three of these—rice, corn, and wheat—make up almost 70 percent of the world's seed crop.

Many of the farmers we spoke with expressed a desire to grow a diversity of crops—beyond simply the diversity required in operating a CSA—or to diversify their production to avoid a crisis if a crop failed due to the seasonal vagaries of disease or weather. The question "What are the principal varieties of crops you grow on your farm?" gave pause to nearly every New England farmer we interviewed. They would often struggle to rattle them off, finishing with a response such as, "And a whole lot more I can't remember right now." Sally of Old Times Farm said that it was the farm owner's intention to grow "stuff that is a little more exotic or unusual, and part of that is just to keep from being bored. He likes to try his hand at growing new things and trying new things." The variety of crops at Old Times Farm was even part of why Sally became interested in working there and has continued to enjoy being there over the past nine years. "I was introduced to a lot of new things that I never would have tried if it weren't for the farm."

When asked about the diversity of crops on Heritage Harvest Farm—a 65-acre non-certified farm using "organic methods" with a 510-member CSA—Claire began by rattling off the less common varieties of crops first. "Right now we have black radishes," she offered, before admitting she couldn't recall most of what they grew. "When a farm is so big and has so much going on, it's really hard to keep track of what isn't your particular niche. But we grow a lot of stuff!" Meghan, a recent college graduate who took a job on Sustainable Harvest Farm—a 130-acre vegetable farm with a 1,500-member CSA—had a particularly hard time listing the number of varieties the farm grew. "We grow over a hundred varieties of tomatoes," she knew for certain, "but for every crop there is going to be at least two varieties."

While John and Katie never explicitly expressed a potent desire to preserve agricultural biodiversity and felt some heirloom varieties were just too fussy for the limited demand, they certainly seemed to be doing their part. Veiled in mesh, four different varieties of cucumber were just getting ready to flower. Once they developed, none of them would look anything like what one could typically find in the grocery store. Likewise, several unusual and colorfully speckled lettuce varieties and a plethora of heirloom tomatoes were being cultivated. Indeed, all of the tomatoes on the farm were heirloom—or traditional, non-hybridized—varieties because customers, including restaurants, were willing to pay more for their superior flavors and textures. As an organic farm—and more important, an organic CSA farm—this type of diversity was a critical component of Scenic View's practices.

As we survey this diversity, it is hard to believe that John's farming days began so inauspiciously or that his family thought his new career was "a joke." The two fields separated by a rustic red barn

spoke to John's farming success. John and Katie had started cultivating these fields only a few years ago, as their customer base was
expanding. As for the barn, construction was started six years ago
to provide space for a commercial kitchen. John wanted to begin
supplementing farm sales with value-added products such as butter
flavored with garlic scapes, the curved flowering stalk of the garlic
plant. Products such as "garlic scape butter" would be a new source
of income for the farm, and a profitable one too, since they would be
more valuable than the vegetables themselves. The undertaking was
given a boost when the farm received a $20,000 government grant
for guaranteeing to preserve five acres of land in agricultural use for
five years.

An attractive yet functional building, the barn was painted the
iconic "barn red" on the outside, while the interior was paneled in
unfinished wide wooden planks. The rough wood grain was a rustic
touch, made even more meaningful by the fact that the timber came
from the woods on the farm. After countless hemlock trees on the property had died from an outbreak of disease, John decided to have a
lumber company come and mill the trees on site. Many of them could
be given a new "life" in the barn.

The barn provided office space, as well as much needed storage for
the farm. It was also outfitted with a used walk-in refrigerator purchased from a local florist. This appliance was a huge step up for the
farm, which had previously stored vegetables in plastic coolers topped
with frozen water bottles.

Despite the fact that the kitchen was not yet functional after six
years of work, the barn provided a new source of revenue for Scenic View. For several years, the farm had sold vegetables at a local
farmers' market (along with sales to restaurants and its CSA operation). However, with the birth of their son, the hassle of market days
became too much for John and Katie.

The struggle to keep up with demanding farmers' market schedules was echoed in our interviews with the other New England farmers. Lucy, a co-owner and farmer at Long Days Farm, which manages
a three-hundred-member CSA and both vegetable and livestock pro-

duction, was noticeably flustered at the market. "It's Wednesday," she quipped, "and I've got to run the stand and manage CSA share pick-ups. I haven't even eaten my lunch yet." It was nearly five o'clock, and after a long market day Lucy was visibly fatigued. With the space provided by the barn, Katie and John at Scenic View were able to set up a small farm stand, which they operated two days a week. While not the most profitable part of the farm's operations, the farm stand serves as an important supplement to their income. In addition, John hopes that people would enjoy buying their food where it is grown. He doubts that most care, but he hopes that they do.

The desire to develop community, to have neighbors and visitors come to the farm for vegetables and good food from the farm kitchen, was a genuine desire for John and Katie and one shared by many such small-scale farmers (DeLind 2003). Certainly the farm kitchen and farm stand were intended to generate new sources of income. They also were meant to allow for a slightly slower-paced marketing scheme and way of life for a growing family that no longer wanted to keep up with hectic days at the area markets. However, they were also intended to bring people and food together in a way our modern industrial system of agriculture has made all but impossible: at the source (Magdoff, Foster, and Buttel 2000). John and Katie's decision to build the commercial kitchen can be seen as an attempt to forge a good match among their need for new revenue, their desire to keep their work on the farm, and their vision of the types of relationships with the community they hoped for. Such desires reflect Scenic View's commitment to "permanent" community-based agriculture. Just as the diversity of the landscape reflected a view of the farm as a living ecological system, John and Katie's desire to share the farm with the community reflects a vision of the farm as a social system. As farmer and essayist Wendell Berry has argued, "Farming by the measure of nature, which is to say the nature of the particular place, means farmers must tend farms that they know and love, farms small enough to know and love, using tools and methods that they know and love, in the company of neighbors they know and love" (1990, 210). The Scenic View owners never waxed quite so philosophical as

Berry. However, the sentiment was certainly there when, with the long-awaited completion of the kitchen, John expressed his hope for the day when people could come to his farm, eat a good meal, and know exactly where it had come from and who had prepared it. The hope, at least, was that an alternative food system could bring about community if people cared enough to know where their food was grown. This is an essential component to the education process that many organic farming advocates argue is essential to long-term, small-scale organic farming success (Donahue 1999).

While many small-scale farmers may desire the type of community John and Katie expressed when they envisioned people coming to the farm to truly connect with where their food had been grown and share a meal, such a vision is often easier said than done. Often, the term "CSA" can be a misnomer. Rather than community supported agriculture, these programs often become community-*funded* agriculture. The CSA is framed "as *an alternative market arrangement rather than a partial alternative to the market economy*" (DeLind 1999, 5; our emphasis). Reflecting on her experiences starting a CSA and on the state of CSAs generally, DeLind writes, "Few are really sharing the burdens of food production or the embodied experience, but are providing a pleasant and thoroughly necessary brand of subscription farming" (6). By supporting Scenic View Farm's CSA program, people demonstrate a concern for knowing where their food was grown. However, the concerns seem "centered in the marketplace and not in the living place" (8).

Sally of Old Times Farm began working on the farm after volunteering as part of her first season's share. When I asked her if other members ever volunteered, she could only laugh. "Nobody does! I was the only one; as far as I know, no one asks." A similar story has played out on Heritage Harvest Farm. When Claire spoke about her experiences working there, she passionately asserted, "I think more people need to volunteer, even if it's just one day, on a farm," to help reverse what she saw as a societal disconnection from where our food comes from. When I asked if CSA members ever volunteered, she replied, "Yeah, I think we just have one." While John hoped to build community, he also acknowledged that most people's concerns for where their food was grown fall far short of a desire to share in the life of a farm like Scenic View.

Despite obvious signs of success, finding new sources of income seemed to be a constant pressure on Scenic View Farm. For example, one afternoon, John discussed with Katie his hopes of possibly bringing more land into cultivation. John hoped that by crossing the line between a small- and medium-sized farm, he might be able to secure for his family a solidly middle-class existence. Katie disagreed. To add another field—to get big—would require more equipment, workers, and time. She was convinced that their take-home income would remain stable in the face of rising costs. John conceded that she might be right.

Collin of Viridian Farm had just closed on the purchase of a new piece of property for his growing farm when we spoke, and although he was obviously excited, he seemed to echo many of Katie's concerns about Scenic View. Worries about "the extreme amount of debt I'm about to go into" were Collin's first thoughts about the new purchase, although he hoped that such "happy debt" would be motivation to do a better job farming.

However, increased agricultural output was not the only way for Scenic View to increase its income. As a result of the farm's natural beauty and quaint charm, John was able to convert two buildings into guesthouses that could be rented out for retreats, conferences, reunions, and weddings. One is is an eight-bedroom home with a swimming pool and dance hall, while the other is a three-bedroom cottage. While the rental business did not require the commitment of running a bed and breakfast—guests were merely renting the property and provided their own services—it still represented a significant investment of time and energy for managing the frequent rental inquiries. Almost every weekend, people were renting the properties, and the houses were booked through till the following year, with only one or two openings left.

John also owned another home, which he rented out as part of his "retirement plan." "There's no easy retirement from farming," John said after getting off the phone with the woman who would clean the house before a new set of tenants moved in. "Somehow, I managed to clean myself up enough to convince a bank to give me a loan to buy the place." Once the house is paid off, John hopes to have another source of income sufficient to support his family in his old age. Finding sources of supplemental income was a necessary reality for Scenic View, and such income helped John and Katie make it as a small farm.

The search for new sources of revenue is, one could argue, a defining characteristic of modern small-scale farming in the United States. This is not a new phenomenon for New England small farmers. The necessity for diversity on the small-scale farm goes back to the colonial period. Take, for example, the description of the homestead of Job and Anna Brooks, found conveniently on a plaque along the Minute Man National Historic Park trail. The Brooks were farmers living in Concord, Massachusetts, in the mid-1700s, but they did not simply farm. No one did. The Brooks family also had a tannery, ran a slaughterhouse, operated a brick kiln, and worked a sawmill. As the plaque states, "Few people in the 18th century were solely farmers. Most men supplemented their income through trades such as shoe making, cabinet making, rope making, or blacksmithing. Many local men worked at the [Brooks's] tannery. Women made dairy products, spun and wove wool and linen cloth, and practiced midwifery. There was a very lively local exchange of goods and services" (National Park Service). To make it as a farmer in New England, one always had to hustle.

As we noted in our brief survey of the state of contemporary American agriculture, despite some increases in the number of small farms over the past decade, agriculture in America remains increasingly bifurcated; a relatively small number of large farms is capturing a greater share of the market, while small farmers scrape by on less. The rich get richer. "Eighty percent of farmers on small acreages have farm incomes below the poverty line, and 59 percent of farms have less than $10,000 in sales annually" (Blatt 2008, 8). Such figures are not specific to the United States: for most of the industrialized world, small family-farm operations rely heavily on off-farm income in order to make farming work (cf. Pritchard, Burch, and Lawrence 2007; Alasia et al. 2009; Oberholtzer 2009; Oberholtzer, Clancy, and Esseks 2010). The reality, for most farmers in John and Katie's position, is that working in agriculture is simply not enough.

What does it mean that farming simply doesn't pay for thousands of small farmers? For John and Katie, it has meant incorporating value-added products into the farming system, building a commercial kitchen, renting guesthouses for supplemental income, and contemplating radically altering their workloads by expanding the amount of land under cultivation. Apart from the last, these options fit with their broader goals and values: a commercial kitchen could help foster community, and renting their guesthouses did not diminish their ability to farm in a principled manner. They were good matches, integrating their economic lives into the lifestyle they saw for themselves. However, if it wasn't making garlic scape butter and renting guesthouses, it would have to be something else. Farming alone not only failed to provide a stable existence, but it wasn't even enough to get by.

If a farm doesn't have a commercial kitchen to make value-added products or a scenic piece of property ideal for weddings and retreats, the quest for supplemental income tends to take farm operators and their spouses off of the farm and into the retail and service economy. According to the USDA Economic Research Service, most farm operators in the United States find work in other natural resource–related industries, like forestry or mining, or in construction and the service sector. Meanwhile, the off-farm employment opportunities for the spouses of farm operators tend to be concentrated in a handful of sectors: education, retail sales, and health care (Weber and Ahearn 2012). The importance of off-farm income ties farm households to the broader economy, often in troubling ways. While the media characterized the farm economy as booming in the midst of the 2008 recession, rural areas were in fact losing jobs faster than the rest of the country (Drabenstott and Moore 2009). As it becomes harder to make ends meet with a farm income, non-farm opportunities are defining what it means to succeed in agriculture.

Farming is increasingly a part-time job. Even in the 2007 USDA Census of Agriculture (2009), which showed an increase in the number of American farms, the greatest gains were in the category of farms with sales of $1,000 a year or less. "In 2004 nonfarm jobs accounted for 91 percent of the income of farm households" (Blatt 2008, 8). Most small farmers are increasingly looking off the farm to pay their bills. Moreover, most are no longer considering farming their vocation. "Of the 956,000 farm operators who indicated that their primary job was off-farm work, 725,000 (76 percent) said that off farm work was now their career choice" (ibid.). John isn't the only small farmer looking for a retirement plan or building a commercial kitchen in pursuit of value-added products; many farmers like him, across the country, are increasingly putting their farms on the back burner.

In the early 1930s, only one in sixteen farmers worked more than two hundred days a year off of the farm; by 2007, nine hundred thousand of them—or approximately one out of five of the nation's principal farm operators—were doing so (Munoz 2010). The money is certainly needed desperately. "Of the 2.2 million farms nationwide, less than half show profit from their farms. The remaining 1.2 million depend on non-farm income to cover farm expenses" (1C). For many, the decision to take a job off the farm is a difficult one. John Mesrobian, a sixty-two-year-old grape farmer in California, wants to farm full time. Although he had hoped to give up his paper-shredding business and devote his time to farming, this hasn't been possible. For the time being, Mesrobian will continue moonlighting on his farm. He cut to the core difficulty of juggling two jobs, saying, "How do you do both and still spend the time you need to with your family? There's no price on that" (ibid.).

For others, the juggling proves to be too much. Such was the fate of another California grape farmer, Dino Petrucci. "Petrucci used to farm grapes in Madera, Calif., and run a catering business but now leases his land for pomegranate production and has scaled down to barbecuing on the weekends" (Munoz 2010). Burning the candle from both ends took a toll on Petrucci, as it undoubtedly does for many farmers like him. In the end, Petrucci needed less tenuous employment; farming didn't fit the bill.

Although all of the New England farmers we spoke to worked full time on their farms, their field laborers did not. Many of the farmworkers we spoke to were recent college graduates who had either failed to find employment in their chosen field or were taking a year off before moving on to other pursuits. While some of these workers—like Elaine, on Pioneer Farm—expressed a long-term interest in agriculture and a desire to return to the farm for the next season, many did not. Meghan of Sustainable Harvest Farm, who was working for only one season before enrolling in graduate school in the fall, described how difficult it was for Sustainable Harvest Farm to coordinate its large labor force: "There is a high turnover from year to year. And although the farm can generally find workers from area colleges, it's hard to have to hire and train new workers every season. So that's definitely one of the farm's biggest challenges." For such farm workers, the seasonal nature of their labor means farming is a temporary pursuit, and the search for a stable career will lead many of them off the farm. Just as for many farmers, for the seasonal workers to whom we spoke farming doesn't seem to pay off in the long run.

Fortunately for John and Katie, the beauty of their farm landscape—and its close proximity to a thriving city center—meant that the scramble for additional income didn't need to drag them off their land and away from the lifestyle aspirations that drew them to farming in the first place. The fact that Scenic View was a family farm, that it was a mixed vegetable farm with a diverse array of vegetables and flowers, and that protected natural woodlands and wetlands separated the non-contiguous fields meant that people were willing to pay to "get away from it all" on the farm. And with the commercial kitchen finally installed in the barn, they are hoping to find new revenue from value-added products. The particular resources available to the farmers at Scenic View made these types of matches feasible. While not every farm can rent guesthouses to urbanites seeking idyllic pastoral getaways, the trends in non-farm income suggest that today's farmers are scrambling to match their economic needs with the lifestyles they desire as best as they can.

Despite John and Katie's avoiding off-farm employment, the quest for new revenue on Scenic View Farm has nevertheless taken its toll. When we talked to John and Katie about finances, there was a sense that the lifestyle they sought through organic agriculture precluded a stable existence. Or was it the opposite? Does attaining the economic security of a modern American middle-class lifestyle necessitate that farmers engage in less economically, socially, and environmentally sustainable practices? If not, how are good matches made among these seemingly incompatible desires? Such questions will be considered in the coming chapters. For now, suffice it to say that the tensions were palpable despite the obvious signs of John and Katie's success.

This discussion is not meant to be a romantic portrayal of their experiences. Far from it. Their agrarian pursuits meant that John and Katie were engaged in a constant struggle to get ahead. Here again we can see the utility of thinking about good matches when interrogating the hard choices made on small organic farms. John and Katie's struggles to get ahead were struggles to integrate economic activity into their lives in ways that both paid the bills and supported the type of lifestyle they endeavored to have. Their success in organic farming—and in their rental business—meant that they could keep farming as their primary occupation. For them, that's what seemed to matter. After all, who lives in a place this beautiful?

five WHO FARMS?

A significant part of what makes the place that is Scenic View Farm, and others like it, is the people who work the land. Like all people, these are complex individuals with rich life histories, the surface of which we will only scratch in the conversations presented here. These are individuals who have overwhelmingly desired a career that would place them out of the office and in the outdoors and who have felt farming was the best way to achieve the lifestyle they desired. They were also individuals who expressed an unwillingness to farm in any way other than organic, either because they knew that the use of chemical pesticides would prevent their full enjoyment of the agrarian lifestyle—such as experiencing the simple pleasure of eating a sun-warmed cherry tomato straight from the vine—or because they did not feel comfortable introducing chemicals into either the environment or the bodies of their customers. These were individuals who worked hard to make organic farming work for them—in other words, to balance their economic needs and aspirations with the host of social and ethical principles of organic farming.

Lifestyle considerations serve as both the primary sources of Scenic View Farm's alternative identity and the primary threats to its organic integrity. This dual reality emerged as a central theme in the stories and practices at Scenic View, and such is likely the case on farms across the United States. Musing on the inherently cultural characteristics of agriculture, Wendell Berry writes, "It is only by understanding the cultural complexity and largeness of the concept of agriculture that we can see the threatening diminishments implied by the term 'agribusiness'" (2002, 285). For Berry, agriculture is distinctly cultural, composed of beliefs, values, and practices. The logic of agrarianism stands radically opposed to the logic of running a conventional business. There is an economic logic to small-scale organic agriculture; agrarianism, however, is not solely an econo-mistic outlook. Instead, it is a response to the many perceived ills of contemporary society, economy, and ecology. As Eric Freyfogle enumerates, "There are the broader anxieties, vaguely understood yet powerfully felt by many, about the declining sense of community; blighted landscapes; the separation of work and leisure; the shoddi-ness of mass-produced goods; the heightened sense of rootlessness and anxiety; the decline of the household economy; the fragmenta-tion of families, neighborhoods, and communities; and the simple lack of fresh air, physical exercise, and the satisfactions of honest, use-ful work" (2001, xvi). However, as Bell remarks about the Iowa farm-ers he studied, "Because of the image of farm life as a refuge from the 'rat race' (with its treadmill imagery), I couldn't help feeling at times that some farm families almost feel betrayed by the reality of their lives" (2004, 71). As we talked with and observed the folks working on small farms like Scenic View, a reflexive spirit of agriculture as part of a broader set of ethical and socio-ecological lifestyle considerations—and the economic challenges a life in the business of small-scale farm-ing brought—clearly emerged as both a central theme in their stories and an ever-present consideration in their practices.

Working on a small organic farm involves spending a lot of time with the same group of folks. It is easy to get to know each other, and it is easy to make friends. Perhaps it has something to do with working together to bring life amid the diverse landscape. Perhaps it is because the choice to farm is so obviously (certainly not solely) a financially difficult one. Whatever the reason, participating in the daily life of organic farming can bring forth rich conversations that often get to the heart of the matter in a very short time.

Our goal was to understand how farmers make sense of the work that they do, in a sector where the work rarely pays the bills yet pro-vides a possibility for achieving sustainable food production and live-lihoods. We sought to explore how individuals who often cast their involvement in organic farming as a lifestyle choice also simulta-neously made sense of it as a job. And we sought to see how these farmers, who often had to make tough choices in order to stay afloat, made sense of these choices in light of their social and environmental commitments. Sometimes, surprisingly, it seemed that such questions did not fit the context. "Why did you get into farming?" for example, proved to be a simple question with a complex, multi-layered answer.

John, whom we have already met in this book, was at first sight a typical farmer. A tall, slender man, in excellent shape at forty years of age, he once described his role on the farm as "moving heavy things and wrecking stuff." Like the stereotypical farmer, John always had something hanging from his mouth, whether it was a blade of grass or an entire baby fennel plant. John is a quiet, kind, and humorous man, often cracking jokes in the fields. John cared about people as much as he did about his crops. John and Katie's employees never grumbled about wages or working conditions. Rather, they spoke overwhelmingly of their generosity. While labor accounts for about 17 percent of the average farm's expenses and up to 40 percent of operating expenses for labor-intensive crops like those grown at Scenic View, 50 percent of John and Katie's expenses were employee wages (Hertz 2014). However, the generosity to which their workers referred extended beyond wages, to the important relationships Katie and John cultivated with their employees. John helped one of his employees move her belongings when she relocated to a new apartment, John and Katie provided another employee with lodging to make it easier for him to work for them, and they hired an acquaintance for as many hours as possible when she lost her full-time job. John invited his employees over for dinner from time to time, and on Fridays he and Katie would provide post-work beer while they socialized with their employees. In fact, one long-term intern suggested John may have been too easy on his workers: if a vegetable bed required tedious hand weeding, he would soon advise his employees to take a break from the arduous task—even if that meant the task remained unfinished for another day.

This type of concern for employee well-being came up in many of our conversations with individuals working on other small farms across New England. When we asked Claire what it meant for Heritage Harvest Farm to use sustainable methods, she didn't merely talk about crop rotation or the avoidance of chemical inputs. Rather, she said, "We are sustainable also because the owner pays fair wages to all of his workers." Likewise, Susanna—an employee and market manager for Happy Hen Pastures, a non-certified organic farm producing pasture-raised eggs and poultry—described the farm's decision not to send its birds off-farm for processing. By and large, this decision was intended to give the farm greater control over the processing of its animals. However, she also described the labor involved on the farm. Happy Hen Pastures employs five local women to hand-butcher all of its poultry. As she described it, this meant the farm was providing valuable, high-skill jobs rather than the assembly-line work characterizing larger processing facilities.

Elaine at Pioneer Farm also described her employment conditions in overwhelmingly positive terms. She had worked on Pioneer Farm for just a few weeks when she was just out of college, but she came back as a permanent employee because "I liked it so much; [my boss is] one of the greatest guys to work for." Leslie Duram, in her study of organic agriculture, *Good Growing* (2005), found similar concerns. In Duram's book, we learn of Phil, an organic vegetable farmer from California who in 2004 was trying to provide competitive wages between $8.25 and $10.00 an hour. Phil also tried to ensure that workers were able to do a variety of tasks on the farm rather than having to spend long, tedious days laboring on the same task. Phil reflected on his treatment of his employees: "Well, the people that work on the farm aren't compensated nearly enough for what they do, but it's as well as we can do" (cited in ibid., 146). Unlike the exploitation of agricultural workers that characterizes so much of the industrial food system, for these farms and farmers achieving a good match between their economic constraints and their personal relationships meant providing sustainable conditions for their employees—conditions that are even perceived by fieldworkers as generous (cf. P. Allen 2004).

Although John of Scenic View fit some of the stereotypes of farmers, like chewing on grass and wearing cowboy hats, he disproved many more. He was an organic farmer, after all. First, he was no provincial. An educated man, John had a college degree and had spent several years overseas after graduating. Second, he was not a conservative. One afternoon over lunch, one of the workers was explaining to John's son that he had been over to her house to watch President Obama get elected. John quickly spoke up. "Isn't it nice to get to say, 'Obama got elected'? It would be awful to have to say we watched McCain get elected. That would have just been too much."

John's quirks are, in fact, characteristic of the organic farming movement. As we saw above, the organic farming movement was not the result of like-minded people agitating for a new, systematic method of agriculture. Rather, organic agriculture—as we now know it—was the product of conservatives and liberals, iconoclasts and hippies, occultists and health fanatics. Moreover, the spectrum of organic practices we have inherited from this decades-long movement are not uniformly applied across all organic farms. And when various organic practices are employed, they do not necessarily carry the same meaning for every farmer. Stereotypes abound concerning organic farms and the farmers who tend them: they are all hippie environmentalists, Luddites, or dastardly snake oil peddlers (see Guthman 1998). The truth of the matter is that there are no one-size-fits-all categories on an organic farm. However, while it may be difficult to define the "typical" organic farmer, some commonalities emerge. For example, current organic farmers are not only younger, more educated, and more likely to be women than their conventional counterparts, but they also have higher levels of environmental concern and less concern for economic profit than their conventional counterparts, or even those farmers who have transitioned out of organic agriculture (Läpple 2012). However, the average age of organic farmers—like that of all U.S. farmers—is rising. According to the U.S. secretary of agriculture, Tom Vilsack, "The average age of the farmer in America today is 57. . . . [In 2007] we had a 30 percent increase in the number of farmers over the age of 75 and a 20 percent decrease in the number of farmers under the age of 25" (Vilsack interviewed in Hansen 2011). Based on the most recent (2012) Census of Agriculture, the average age of the American farmer today has risen to over 58 years (USDA 2014).

Despite the fact that the United States has lost 20 percent of its farmers under the age of twenty-five, growing numbers of young college graduates are trying their hand at sustainable agriculture (see Weise 2009). Although John did not start farming at such a young age, a college education is one similarity John shares with many organic farmers. In fact, 81 percent of organic farmers surveyed in 2004 indicated they had had at least some college (Walz 2004, 101). Much like John, 26 percent of organic farmers have a bachelor's degree. Moreover, one-fifth of U.S. organic farmers hold graduate degrees (ibid.). In 2004, 49 percent of organic farmers began their farming careers in organic agriculture, as opposed to the 51 percent who transitioned from conventional practices. Such data would suggest that John was much more "typical" than conversations revealed. A significant number of organic farmers are middle-aged, college-educated males who began their farming careers in organic agriculture. It is worth noting that all but two of the farmers and farmworkers with whom we spoke across New England had college degrees. The two that didn't had grown up working in agriculture.

Determining how typical John's beliefs are—about farming, the environment, or politics for that matter—is another issue entirely. For example, very little is known about where organic farmers fall along the political spectrum since USDA surveys of organic farmers do not ask (Sayre 2011). It is nearly impossible to say whether John's sigh of relief that President Obama was elected in 2008 was part of a collective sigh across the organic farming movement. Most likely it was not. According to Laura Sayre, "It's become a cliché to point out that 'organic farming is not just hippies in Birkenstocks,' but this doesn't convey the half of it. Organic farming is hippies in Birkenstocks, hippies in business suits, born-again Christians in Birkenstocks, everything in between, and a whole lot more besides" (39).

Some involved in the organic movement believe that the common stereotypes of liberal organic farmers are fundamentally misguided.

For example, one Iowa organic certifier offered the following characterization of the organic farming community: "Organic farming is
conservative small-time rural farmers making food for white liberal
yuppie and hippie types" (cited in Sayre 2011, 39). And while the concentration of organic farms is in the nation's "blue" states, there is no
way to know whether this reflects the political affiliations of the organic farmers or merely the organic consumers that these farmers rely
upon to market their food. At the very least, the political beliefs of the
farmers in the organic community are anything but homogenous (41).

As with political affiliation, research has shown just how difficult
it can be to describe "the organic farmer" as a theoretical ideal-type
representing a unified environmental or agricultural perspective.
While thinking about organic farmers as a homogenous group is impractical, it may not even be helpful for understanding alternative
agriculture in New England or the United States. For example, William Lockeretz (1997) has found that organic farmers must be understood as responding individually to their particular social, economic,
and personal contexts. The particular intersection of a farmer's biography, history, and social context serves to produce local expressions
of organic practice. Likewise, Leslie Duram has argued that organic
farmers seem to articulate concerns about economic, political, social, and environmental structures. However, each farmer responds
to these structures differently; there is no single organic approach.
Rather, "It is necessary to understand structures from the viewpoint
of individual farmers who mediate them to help create our rural
landscape" (2000, 47).

Nevertheless, research has shown that organic farmers may, at the
very least, share a common disposition or outlook. For example, in a
study of both conventional and organic farmers in Colorado, Duram
found that "reactive attitudes and characteristics seem to constrain
farmers to maintaining their conventional methods of production.
On the other hand, the adoption of alternative methods is accomplished by farmers whose proactive attitudes drive them to adapt to
current uncertainties in U.S. agriculture by trying alternative methods of production" (1997, 212). This characterization seemed to fit

many of John and Katie's practices. For example, their decision to start a CSA rather than utilize existing market structures, farm in ways that worked with nature rather than conquering it, and promote a diverse farm landscape rather than a monoculture all align with what Duram classifies as a "proactive" farming outlook (208). Yet a proactive attitude can be beneficial in any number of careers. To understand why John and Katie were farming at all, we have to look elsewhere.

Figuring out exactly why John responded to economic, political, social, and environmental conditions by trying to envision a new agrarian future for himself, his family, and even his community proved particularly difficult. At first, it was hard to pin down why John was a farmer at all. Regardless of how he responded when asked, it always seemed that there was more to the answer. In light of the unpredictable finances and limited economic security that have driven so many out of small-scale farming, the choice of a career in agriculture seemed to necessitate a compelling rationale.

According to Bill McKibben, "By 1980 there were so few farmers left in the country that the Census Bureau no longer bothered to list farming as one of the occupations you could check off on its form" (2007, 55). Less than 1 percent of Americans currently claim farming as an occupation (Economic Research Service 2015). Most Americans never interact with a farmer, except for the few who buy produce at farmers' markets, and then it is only for a few minutes. As a result, how anyone ended up farming these days is almost a lost question. Why does a college-educated, well-traveled individual like John—with the opportunity to live a thoroughly middle-class urban or suburban life—choose to live a working-class existence on a small farm?

Cultivating an Agrarian Vocation: Good Matches and Farmer Identity

Apart from describing how he wanted to get back to his family's land in order to enjoy the sheer beauty of the place, John gave various answers to our questions. He also mentioned a love for food, a category into which farming seemed to fit. Finally, at a barbeque that John and Katie put on for a group of visitors and family at the barn, and to celebrate the long-awaited completion of the kitchen, some answers emerged in more detail about why John wanted to be a farmer.

When asked for a fourth time why he farmed, John explained that in the end, he was just an outdoors person. He then told a short story to illustrate the point: "When I was a kid, I used to do a lot of fishing in the summers. I sort of realized that all of that fishing was just a way of being outside all the time." In a way, the illustration was satisfying. It may not have explained exactly why John was a farmer and not a construction worker or a park ranger. However, combined with the beautiful land, his love for food, and a concern about conventional food production, his occupation made sense. It was a way of matching his economic life with a view of himself as a whole person, signaling through his career the vision of himself he sought to cultivate. Taken together, John's answers speak volumes. John's choice to become an organic farmer was less about what he wanted to do than a reflection of who he wanted to be.

Although John's fishing anecdote and reported love for the outdoors might not initially satisfy one's curiosity about why someone with his background would choose farming, John was articulating a very powerful ideological commitment. While "being an outdoors person" may not seem ideological, it reflects a commitment to a particular lifestyle. Indeed, the cultivation of a particular type of self is at the heart of Michael Bell's (2004) phenomenological approach to alternative farming. Similarly, Lyons and Lawrence (2001) have identified such lifestyle choices as important factors in people's individual decisions to become organic farmers. They found that many organic farmers were highly qualified individuals with higher-education degrees looking for a new way of life. For example, one New Zealand orchardist reported, "Living in Sydney and just being surrounded by buildings, we thought there must be more. And I guess for me, I was interested in organics over conventional because while living in Sydney I had become quite interested in environmental issues" (cited in ibid., 79).

The search for "something more" than a desk job certainly came across in our interactions with John. It seemed as though he had always wanted something different from the middle-class nine-to-five job. This desire for a richer, more meaningful life connected with nature was not limited to John. Every person at Scenic View Farm sounded a similar refrain: farming was a lifestyle choice.

Katie, who was a quiet and friendly woman, had been married to John for the past five years. Farming had become a good fit for her, reflecting her values surrounding the issue of work. She too spoke about loving to be outside: "You can't beat farming when the weather is good," she once remarked. Katie's strengths seem to complement John's in many ways. For example, although Katie believed that farmers who claimed they did not get certified as organic because of the paperwork involved were using the bureaucratic wrangling as a convenient excuse, John conceded that he probably couldn't keep up with it without her. Although Katie was in the fields as much as anyone else, and despite her dislike for office work, she appeared to have a knack for the business side of farming. Katie kept most of the records required to maintain the farm's organic certification.

Despite Katie's skepticism of farmers who claim the paperwork kept them from becoming certified, keeping up with such paperwork is critical to certification and can be an onerous task for many farmers (Guthman 2004b). Losing organic farmers for any reason, such as the burdens of certification requirements, might be a serious obstacle to maintaining the small-scale organic farming sector. In a study of California farmers who decided to discontinue their organic certification, 50 percent cited the regulatory paperwork and record keeping as serious concerns (Sierra et al. 2008, 34). It takes certain relational practices, economic matches, and beliefs to make the current system of organic certification workable on small farms—many of which we will describe in the following chapter. Katie's willingness—and ability—to manage paperwork was a critical part of how Scenic View maintained its organic certification.

Apart from keeping a handle on the farm's record keeping, Katie also kept a handle on John's admittedly short attention span. During the cold and wet weather of the early summer, when John was itching to complete new tasks, Katie would remind him not to handle the plants too much or to stress them through pruning, in order to prevent the spread of disease. Despite being good at such managerial activities, she did not enjoy the stress that came along with them. On one particularly busy day, when the farm had several volunteers to coordinate and a number of large orders to fill for restaurants, the stress was written all over Katie's face. Once the orders had been successfully picked, washed, and packed for delivery, her characteristic smile returned.

Also working on the farm was Mark. He was a funny man, always telling jokes and teasing the workers from behind a pair of dark sunglasses. He was never without sunglasses, hating to have the sun in his eyes, as well as hating the thought of the damage it was causing the eyes of those who did without a pair. His language was often coarse, and he would frequently follow up a string of swearing with the question, "Did you get that? Put that in your book!" Like John, Mark also seemed youthful for a man of forty. The same age as John, they have been friends since the second grade.

That friendship formed the basis of Mark's farming career. Shortly after John had begun farming, he had a falling out with a farmer who had been working for him. Mark was in between jobs at the time and came to the farm to fill in for the season. He ended up staying at the farm when he fell in love with the lifestyle, especially working out of doors. Mark isn't just a laborer at Scenic View Farm; he is a friend and partner, reflecting the strong values of community present in the individuals there. While Mark enjoys farming, he enjoys managing the guesthouse rentals that supplement the farm's income much more. Often, while he rode on the tractor cultivating the fields—another of his frequent jobs on the farm—he would suddenly stop, mid-row, to answer the phone.

It seemed as though Mark was always on the phone with renters, and he was obviously good with them, allaying their fears about the rustic nature of the accommodations and arranging tours of the facilities. However, when customers would try to arrange a visit, he would always interject over the phone, "But I'm also a farmer, so that time of day isn't really good for me." Despite not originally intending to stay at Scenic View, Mark seems to have adapted to the life of a farmer and events manager.

Pat is another worker on the farm, and she is fondly known for her weather predictions. Pat explained her lifelong love of the weather: "I just am fascinated by it. It's just out of our control, you know? It happens whether you like it or not." One of the other workers chimed in, arguing that the weather was no longer beyond human control, with the seeding of clouds and climate change. Pat laughed as she conceded, "Well, most of the time it's in God's hands."

Along with being the selected weather expert on the farm, Pat had been employed on the farm longer than anyone else, even John. Since graduating from high school, Pat has worked for John's family for the past twenty years. When she finished school, she didn't know what she wanted to do, and both her father and sister worked on John's family's horse farm (the same one on which John worked over his childhood summers). Pat asked her family to get her a job, and she has been working as a groundskeeper ever since. When John's farming operation started to grow in scale, she began working with him a couple of days a week. Like everyone else, she has stayed with the job because of a love of the outdoors. "I guess you could say I got bit by the outdoor bug," she explained one morning as we harvested beets together.

In fact, all of the farmers and field workers with whom we spoke from other farms in the region seemed to have been bitten by the same "outdoor bug," and they saw farming as, above all, a lifestyle choice. In every interview, the individuals mentioned that they enjoyed working outdoors. Molly, the farm manager of Grateful Harvest Farm—a thirteen-acre, certified-organic vegetable farm with additional non-certified grazing land—came to farming with little previous experience, deadheading day lilies with her mother as a child and urban gardening as a young adult. However, the lifestyle she pursued had an almost lyrical quality: "I love filling my hands each day with dew-touched heads of lettuce, dirt that ingrains all wrinkles, oil-rich handles, tomato suckers, imperfect eggs, infant weeds, mature roots, and the bridge of a bull's nose." A deep desire to be connected to the natural world and its cycles motivated Molly to begin working two and a half years ago on the farm she now manages.

Jane, the daughter of the owners of True Friends Farms who has worked on the farm both as a field intern and a market salesperson, explained: "I just can't sit around at a desk not doing anything. I need to be doing something. . . . I just love going out in the fields and picking some vegetables and eating them. I just love it. It's something I can see myself doing for the rest of my life." Likewise, Collin of Viridian Farms described how he became interested in food after going to college and having to start cooking for himself. On top of that, he had always loved hiking and being outdoors. As he described it, "I began to realize that I could potentially start working toward a career that both got me outside a lot and intersected with nutrition and health ideas, as well as environmental and sustainability ideas, as well as entrepreneurial ideas, and that just really fit with the type of lifestyle I wanted." Sean, who works on his family's farm, Verdant Acres Orchards—a non-organic, 250-acre farm and orchard that sells primarily at farmers' markets and an on-site farm stand—did not attempt to romanticize the work associated with a life in farming. "What's it like? It's hard work. Hot summer days. You're out working at, like, six in the morning, picking and harvesting and planting and tilling, and you're not done until like seven, eight o'clock at night. You're starting early, and you get home at dark. It's tough, real tough work. But I like the work; I enjoy it."

Such lifestyle orientations have been recognized among farmers—organic or otherwise—on farms across the United States. Take Wendell, a farmer described in Michael Bell's *Farming for Us All*. Wendell beautifully described the lifestyle: " 'It's being able to appreciate the night,' he said finally. 'I've gotten to where I like to appreciate nature. . . . I like the simpler things in life. I like farming. I like the theory of getting up some day and you don't have to go to town. . . .

You can say the heck with it. I'll just stay out here and not have to worry about it. I guess maybe it's a different upbringing than the city and its rat race'" (2004, 37).

As Bell explains, the pastoral ideals of American farmers may have to do with something more than being "bitten by the outdoor bug." Rather, agrarianism holds a special place in the American cultural imagination. He suggests, "The hold of farming on our imagination goes well beyond the economic [importance of agriculture]. It is also a matter of the cultural associations we make with farming as a part of nature, an act of simplicity, of honesty, of certainty, however romanticized and empirically inadequate these associations may be" (2004, 36). In Bell's words, the phenomenology of farming is "this taking for granted what you know works, even when you think it might not" (14). Through this approach, Bell pushes beyond the political economy approach, which dominates so much of the theorization of organic, and pays attention to lifestyle concerns and the politics of identity wrapped up in alternative farming. We also pay attention to the lived experience of farming and the ways in which that experience comes to make sense for organic farmers—so much so that it becomes taken for granted. On Scenic View Farm—as well as on countless other small farms across the region and across the United States, organically certified or otherwise—farmers are overwhelmingly seeking a lifestyle viewed as radically opposed to that of the majority of Americans. Through the ways they match the business of farming to their personal lives, they are seeking to mark their own economic, social, and personal lives as distinct from the types of work and leisure pursued by others in the contemporary economy.

Everyone else working on Scenic View farm had been there for less time than those we have already introduced. Maggie, for example, has worked at Scenic View for three seasons. A young woman, she became interested in sustainable agriculture after becoming "disillusioned with the world" during college. In light of all of the world's problems, she sees local agriculture as a practical step in the right direction toward an organic future. For Maggie, the choice to farm organically is an ethical one. She has worked at two farms besides

Figure 5. Life's Simple Pleasures: Husk cherries straight from the fields at Sustainable Harvest Farm. (Photo courtesy of Connor Fitzmaurice.)

Scenic View and is involved in the same regional organic farming association that helped John get started as a new organic farmer.

Despite her interest in organic farming, Maggie didn't see herself as a career farmer. At the time of our interview, she was attempting to rent her apartment in town and move to a suburb of Boston to pursue a new teaching position at a Montessori school. If she were to leave, she would be missed greatly. Maggie was a hard worker and was always smiling. Once after a long, hot day in the fields, she drove into town to buy ice cream for everyone at the farm. While Maggie was a generous individual, such actions were indicative of a general atmosphere of friendship and caring present on the farm.

Several of the farmworkers with whom we spoke from other New England farms echoed Maggie's attitude at Scenic View. Claire was working at Heritage Harvest Farm but was also working on a degree to become a certified health coach. While she enjoyed working on the farm, it was merely a temporary pursuit. Likewise, Meghan at Sustainable Harvest Farm was working in agriculture as a temporary break before enrolling in a graduate program. Yet, like Maggie, they saw working on a farm as important—as a practical step they could take to do something positive in the world. As Meghan at Sustainable Harvest described it, "After spending five years at school working in front of a computer, and before going back to spending the rest of my life in front of a computer, I wanted to do something real—something with my body. . . . In farming, you sometimes need to use your head, but it's mostly your body." And as Mary from True Friends Farm put it, farming was a good break as she was deciding what she wanted to do after graduating because it produced a tangible, positive result. She enjoyed the feeling that she was "doing something valuable."

Back on Scenic View Farm, Joey, a young man, had been an intern for the past two summers. Like Maggie, he had become interested in organic agriculture while at college. However, unlike Maggie, he dropped out of school to pursue agriculture after becoming disillusioned with what he perceived as "the sale of degrees." Learning could happen anywhere, and for him it had happened better in the fields than it in the classroom. A pleasant person, Joey was quiet at first but would talk at length when interesting topics—such as sustainability—were brought up.

Joey was very interested in farming and was hoping to start his own farm or nursery someday. A local, he had looked John up on the Internet and e-mailed him about working to gain some practical farming experience. The previous winter he had volunteered on a farm in Costa Rica to continue learning during the New England off-season. As a result of Joey's interest in farming, John tried to consult him when decisions had to be made, not only getting his opinions, but also helping him to develop as a farmer.

Maggie and Joey probably fit the counterculture stereotype of organic farming more than anyone else at Scenic View. Maggie had a college degree, yet she chose to farm—at least for the time being. And Joey was so interested in getting his hands dirty that he had left school altogether. The motivations behind some of these individuals' decisions to farm have already been mentioned. Other motivations behind these decisions would be surprising and will be given greater attention below. Across New England—and across America—many recent college graduates are trading suits and ledgers for tractors and seed drills.

Breaking from the decades-long trajectory of fewer, graying farmers, many young farmers are new converts to agriculture with no family history of farming. According to a survey conducted by the National Young Farmers Coalition, 80 percent of the nation's young organic farmers did not grow up farming (Charles 2011). Signs of such changes abound. For example, in November 2009 the Stone Barns Center for Food and Agriculture in New York hosted a conference for new farmers, expecting a modest turnout of 40–50 people. On the day of the conference the center was shocked to have 170 participants (Wiese 2009). In Vermont, the Center for an Agricultural Economy hosted a "young farmers mixer." It, too, expected only a modest turnout but found itself hosting 150 people, coming from as far away as New York (ibid.).

The impact of these new farmers on the average age of farmers will not be known until at least the next agricultural census. However, it would certainly seem that organic farming has captured a new generation's imagination, much like the counterculture movement before it. As Dan Charles reports on NPR's food blog, "The Salt," "Growing vegetables has never, in recent memory, been quite so cool, or so attractive to the young and well-educated. Waves of them, perhaps to the anguish of their parents, are migrating into the rural countryside, perusing seed catalogs and learning the finer points of organic weed control" (Charles 2011). Most of these young people are motivated by concerns we heard time and again from the people at Scenic View Farm: the desire for a lifestyle radically different from the one offered by corporate America and supported by the rising popularity of local organic products (Ramde 2011).

Despite the allure of organic agriculture, it is a difficult world for young farmers to enter. The biggest barrier is the most critical: land. Between 2000 and 2010, the average cost of farmland in the United States doubled, from $1,090 to $2,140 per acre (Raftery 2011). In New England, as a result of the demand for real estate, agricultural land values far exceed national averages, making the region home to the most expensive agricultural land in the country. In 2011, the average value of cropland in the region was $7,040 an acre. However, some New England states have land values significantly higher than the regional average. Farm real estate in Rhode Island averages $13,000 per acre, and in Massachusetts and Connecticut it is close behind, with averages of $11,000 an acre or more (National Agricultural Statistics Service 2011).

Other challenges are less obvious. As a result of the steady consolidation of agriculture, with fewer farmers cultivating larger expanses of land and more agro-industrial firms than farm families cultivating those that remain, many of the areas where new farmers can afford land are rather forgotten places. There is a very real challenge of isolation confronting aspiring farmers from cities and suburbs. For some, the isolation is just the price one pays for a beautiful farm and a chance to get one's hands dirty. For example, Luke Deikis of Quincy Farm in upstate New York reflected, "We are in a wonderful spot in rural America, even if there aren't hip bars with good beer on tap" (cited in Raftery 2011, 25). For others, the lack of amenities—and people—takes its toll. Emily Oakley, who moved from California to Oklahoma with her husband in search of cheap farmland, says she has seen the isolation ruin marriages. Describing her own situation, Oakley admitted, "It was just the two of us, every job we did together. . . . It's intense. We would gladly trade a little competition for more community and collaboration" (ibid.).

In an attempt to mitigate at least some of these challenges, particularly the lack of experience among new converts to farming, the USDA is taking action. Secretary of Agriculture Tom Vilsack, for example, called for the recruitment of one hundred thousand new farmers in the next several years, and he has called on Congress to take concrete steps to support that goal (Ramde 2011). In the 2008 Farm Bill, a program was established for training new farmers, and $18 million was allocated to colleges and extension programs to educate individuals interested in pursuing farming (Raftery 2011). The USDA even has boots on the ground recruiting. In the spring of 2011, Deputy Secretary Kathleen Merrigan—known as a "tireless advocate for local food production and marketing, organic food and women farmers" (Vanac 2013)—toured colleges and cities across the country to encourage young people to consider farming as a viable career (Raftery 2011).

Apart from the government, other organizations are exposing young people to alternative agriculture and supporting those who decide to pursue an agricultural vocation. On college campuses across the country, students are demanding sustainable options. In a dramatic change, "nutritionally wired students—many raised on Whole Foods diets at home—are pushing campus dining standards to be measured more by the food's origin, not its volume" (Horovitz 2006, 1B). Recognizing that students who want sustainable food options will go elsewhere to procure them "has moved the $4.6 billion college food service industry to respond with new ways of operating that include relying more on nearby farmers for staples and produce, and serving more organic foods" (ibid.). Reflecting on such trends, Don McQuarrie, executive director of dining services at Yale University, remarked, "Ten years ago, it was OK to get any tomato. Now, we want to be able to tell the students who grew it and where" (ibid.). For those who decide to commit to sustainable agriculture as a career, organizations like the National Young Farmers Coalition and the

Greenhorns have emerged to support their efforts and bring attention to their struggles.

Joey and Maggie were a part of such emerging trends. While the full effect of the emerging youths' engagement with America's agricultural system remains uncertain, the extent of its impact on Scenic View Farm is clear. The farm has apprenticed several interns over the years, although Joey and Maggie emerged from college already connected to the sustainable agriculture movement and committed to organic farming practices. They coveted the lifestyle and ethics of organic agriculture and aspired to it. Defying decades of stereotypes about young people itching to "get off the farm," these young people were willing to put their college credentials aside for what they saw as honest hard work with a purpose. As Katie remarked one afternoon as we were weeding, Scenic View has yet to rely on immigrant labor—there have always been young people eager to work and learn.

Finally, there was Donna, the most recent addition to the workforce at Scenic View. The year 2009 was her first season working and her first experience farming. Trained as an archaeologist and a professional cook, Donna was an interesting individual. One day in early June, she uncovered an eighteenth-century pipe stem while pulling weeds in the field, and it proved to be a good conversation starter. Donna has known John and Katie for a number of years through a food cooperative that she helped to form. When the coop lost its original location, John and Katie offered the farm as a distribution site. Then, when Donna lost her job earlier in the year, John and Katie offered the farm once again.

In addition to needing a job, Donna's values made her a perfect fit at the farm—especially her love of food. A volunteer writer for a local food magazine, Donna was often coming up with recipes at the farm. One afternoon, for example, she picked a pound of purslane (*Portulaca oleracea*, an edible weed) to experiment on for possible recipes. The same spirit of friendship was evident in Donna as in everyone else on the farm. She would offer tea from her thermos, which she brought to the farm even in the July heat. Counterintuitively, she explained, a hot drink could cool one off in the summer heat.

Clearly farming was not merely a job for anyone at Scenic View; the farm was a community and a social statement. Some people there were longtime friends; others knew of each other through memberships in the same organizations. Regardless of the connections, they all came together for a common purpose to work toward a common goal. For each of them, the practices of organic farming at Scenic View allowed for viable intersections of the economic, environmental, and social relationships in their lives.

The common purpose was organic farming; the goal was envisioning a way of life that was commonly perceived as impossible to achieve in more typical middle-class occupations—or even in the conventional agricultural system. These were people who wanted to live their lives with their hands in the cool soil of spring, their faces bathed in the warm summer sun, and their eyes opened to the bounty of autumn. These were people who wanted to farm with the environment they loved, rather than battle against it. Organic farming was a good match for each of them, albeit for a variety of reasons, and, critically, the types of matches they forged between their economic needs and social and environmental relationships offered hope for a more sustainable, organic future. These were people who wanted to be "good farmers."

Paul Stock has developed the concept of the "good farmer" as a way of understanding how the moral concerns of farmers shape agricultural practices and extend ethical considerations to both consumers and the environment. Stock has argued that "family organic farmers have taken a moral stance through their actions as good farmers to create and sustain good soil, healthy food, strong families and strong communities" (2007, 96). William Major (2011) explains it as agrarianism, as a way of life that values relationships and healthy communities, informed by the "grounded vision" of the organic movement and its visionaries that emerged in the 1920s and '30s. Like John, many organic farmers were not always involved in agriculture but were once concerned consumers living on the other side of the agrarian relationship. By becoming farmers these individuals are able to translate their ethical consumption patterns into ethical production. "Family organic farmers recognise the moral right of consumers to purchase healthy food while at the same time they recognise the moral necessity of protecting the land they farm, because at one time many of them were solely consumers themselves." Consequently, "Organic farmers can demonstrate they care not only for individuals but also for the place in which they are rooted—the land, the community, their families and themselves" (Stock 2007, 96). Such moral concerns about relationships among themselves, their land, and their customers are not at odds with a desire or need to make a living. Instead, the pursuit of being "good farmers" requires "good matches": careful, concerted efforts to balance the economic, moral, and sentimental aspects of farming practices in ways that make sense for each farm and farmer.

The objective of being "good farmers" was self-consciously pursued on Scenic View Farm, but it was a pursuit that did not fit the stereotype of alternative agriculture. For example, Maggie suggested that "environmentalism" had little to do with what went on in the fields. Sure, everyone was concerned with the environment, but she didn't see it ideologically—that is, as an "-ism." Maggie suggested that it was not ideology but ethics at work on the farm—John, Katie, and the others had a desire to do what was right—to be good people and, by extension, "good farmers." Maggie was keenly aware that being a "good farmer" was above all a lifestyle choice. It was also a concept Sally of Old Times Farm perceived. When asked whether or not the farm's business pressures ever changed the approach of the farmer for whom she had worked over the past decade, she strongly asserted, "Change the way he farms? Oh, no. No, no, no, no, no. He knows he's never going to get rich doing this. For him, what he does is a vocation. No. Not a vocation. A lifestyle." Such lifestyle choices mean hard decisions must be made that might mean losing out on some financial gains.

John and Katie, like many others, have even chosen to avoid Whole Foods, the organic grocery giant that in recent years began highlighting its commitment to support small-scale local farmers. (It has been described by some commentators as an attempt to bolster its image and remain one step ahead of Walmart's entry into the organic market [see Ness 2006].) In 2006 Whole Foods began requiring that each of its retail locations "buy 'out the back-door' from at least four individual farmers" (1F). However, Katie shared that she and John would not consider selling to Whole Foods, or any other retail location, after hearing "horror stories" from friends in the local farming community who had harvested crops for Whole Foods only to have them rejected for cosmetic defects or insufficient quantities. For John and Katie, cosmetic defects from bugs and disease are just part of organic farming, and Katie suggested they were neither capable of nor comfortable producing at the scale that would be required to meet a grocery store's demand on top of their CSA commitments.

When Sally discussed whether Old Times Farm sold any of its produce wholesale to places like Whole Foods, she responded with the same "Oh no, no, no" she had offered when reflecting upon whether her boss had changed how he farmed over the past decade. "Oh, we're too small. Far too small." In fact, unlike many of the farmers in our study—who were pursuing growth on their farms even while eschewing conventional markets—Old Times Farm had made the conscious decision to reduce its scale from a sixty-five-member CSA to a mere twenty-two members. As Sally explained, it became too much to manage the farm with the demands of so many consumers—many of whom did not seem to understand the fact that CSAs involve the risk of an entire season of Swiss chard if other crops failed to thrive. Rather than conventionalizing and intensifying, Old Times Farm sought to operate at the smallest size feasible that was still economically sustainable.

Even farms that participated in wholesale business typically managed in a very different manner from conventional farms. Nearly all of the farms in our study limited their wholesale business to restaurants. However, most described it as an endeavor they undertook on their own terms. Before working at Scenic View, both Katie and Mark had worked at restaurants in the region interested in farm-to-table fare. As a result of the connections they brought with them to Scenic View, the farm had consistent restaurant sales. However, these remained a small share of the farm's business, accounting for about 18 percent of the farm's revenue (or about $14,000 of net sales). Collin of Viridian Farms described reducing the number of his restaurant clients to only those that could place aggressive enough orders to warrant the more meticulous harvesting, packaging, and transportation that restaurants require. Other restaurants were simply encouraged to shop at the farmers' market at which Viridian sells.

Sally at Old Times Farm admitted her farm sold to a handful of restaurants, "but only because it's convenient for us to do so. They are in the area and are kind of high-end places, places that care about local, sustainable food." Pioneer Farm had recently begun selling wholesale to a regional food hub—essentially a distributor for small local farms. However, it did so only because the sales were guaranteed—unlike at the farmers' markets—and it valued the fact that the farm's name remained attached to the produce when it reached its end consumer, be it a restaurant or a retail store. While Elaine acknowledged that the prices Pioneer Farm received were lower, she explained that in conjunction with the farm's CSA and market sales, it was sustainable and had an added level of security—preventing the late-season scrambling for buyers John noted was absent on Scenic View Farm. Still others described wholesaling as a last resort for surplus crops. Matthew at Peaceful Valley Orchards commented, "We only wholesale when we are long on a particular crop, like apples." Otherwise, Scenic View Farm and the other small-scale CSA farmers with whom we spoke across the region eschewed conventional markets in favor of direct-to-consumer marketing.

The USDA's definition of organic farming practices—which, as we have seen, has come to be based on allowable and excluded agricultural inputs—obscures the lifestyle concerns central to the visions of organic agriculture we have observed. However, the ways in which theories of conventionalization and bifurcation parse out "conventionalized" and "movement" actors within the organic sector suggest that to truly be an organic farmer—a "movement" farmer—one must be concerned with the type of "–isms" Maggie spoke of. By bringing in the language of economic sociology to discussions of organic farming, we can understand the ways lifestyle matters in the experiences of organic agriculture, where money and morals often intersect. The farmers at Scenic View sought a particular way of life through organic farming, one that matched the economic, social, and environmental relationships in their lives in ways they could feel good about.

Scenic View Farm was the place where one small group of alternative organic farmers was rooted—where they sought to live a lifestyle that allowed them to positively impact the land, their community, their families, their friends, and themselves. But how does "good farming" work in practice on a small organic farm? Day by day, how do the practical and economic necessities of small-farm life work in concert with ideologies and ethics to shape what good farming can and cannot look like? A life in organic farming was a good match for each of the individuals at Scenic View Farm for a variety of reasons, but, we would argue, the day in and day out practices of organic farming require similar levels of negotiation among economic, social, and environmental concerns. However, "good matches depend on the stock of meanings, markers, and practices actually available in the local milieu" (Zelizer 2006, 307). Accordingly, we must now turn to how these farmers individually negotiate larger structural conditions—economic, political, social, and environmental—in their daily efforts to live lives that they feel are personally fulfilling, socially valuable, and environmentally responsible.

Scenic View Farm is often a flurry of activity. At the height of the growing season, the farm must prepare to feed one hundred families through CSA orders. The CSA experience perhaps best captures how the promise of alternative agriculture works itself out in both farming and community life. Much of the time, Scenic View is filled with the peacefulness and calm of the farm fields and surrounding landscape. Much of the time it is filled with exhaustion as well, as farmworkers hunch over rows of beets and carrots and squat to pull up weeds. The CSA, however, has brought a new set of exhausting tasks and pleasures to farm life.

A light fog blanketed the fields one Wednesday morning in June 2009, and the ever-present birdsongs were filling the sunlit air when numerous new faces could be seen working diligently to establish a makeshift assembly line. Wednesday morning was when Scenic View Farm prepared its weekly CSA orders. The process of running a CSA brings a new dimension to any small-scale farm. In their study of CSA farmers and consumers, Thompson and Coskuner-Balli describe the ways CSA operations allow consumers access to the backstage work of farming, even if it is but a glimpse. They explain that the consumer participants in their study "effusively praised CSA events, such as potlucks, watermelon tasting events and farm tours, for rectifying the feelings of emotional detachment and sheer ignorance wrought by the separation of food production and consumption" (2007b, 285). Even more, working members of CSA farming operations experienced a sense of enchantment through their involvement in the life of the farm. In these ways, CSA participation brings consumers and producers together to create "market-mediated communal connections" (276).

The new faces at Scenic View were friends and neighbors, all women, who came to the farm each Wednesday morning to volunteer in packing brown bags for the CSA. From inside the barn and out to the back porch, all of the vegetables that had been harvested the day before were laid out at stations. Some individuals were assigned two vegetables to pack and stood in between the piles in front of a few stacked milk crates. Others were assigned to bring out fresh trays of produce from the walk-in refrigerator when supplies got low. Others still were assigned to carry the filled bags into the farm stand in the front of the barn, where they would be picked up in the afternoon.

Before the process began, the regular employees on the farm hurried under John's direction to harvest the last of the vegetables. The day's volunteers chatted while they counted the vegetables laid out on the trays to ensure there were enough. Basins were filled with water for rinsing, while the more delicate vegetables for the week's orders were picked to ensure they were received fresh. Surveying a pile of turnips, Katie decided some of the bunches we had prepared the day before looked skimpy, so more were picked, briskly rinsed clean, and slipped in to supplement the more meager allotments. After the last-minute preparations were completed, it was finally time.

"Everybody ready," a woman called out over the cheerful din of workers and volunteers chatting along the line. At that, the crisp sound of paper bags opening could be heard throughout the barn. Carefully, each bag was filled with each of the items harvested for that week's share, precariously balanced on the stacked crates. At the end of the line, the contents were topped with a bouquet of lavender flowers and then brought over to the farm stand. The pace of the work was brisk, but the barn was filled with laughter, the banter of volunteers, and the forging of new personal connections. Joey's mother was volunteering for the first time that week and was warmly welcomed by all of the other volunteers and workers as they settled into getting the work of the CSA done. These may have been "market-mediated communal connections," but the communal aspect of the CSA model became palpable among these volunteers (Thompson and Coskuner-Balli 2007b, 276).

It was not long before the same woman who had heralded the start of the packing process called out again, "Last one for pickup." After the last bag was taken to the farm stand, the group reassembled and began again, this time loading the bags into the back of a pickup truck for local delivery. The whole process was repeated once more, and the bags were loaded into John's minivan for another drop-off location. By shortly after ten o'clock, the flurry of work was done, and Katie sent the day's helpers off with a heartfelt thank you and with whatever vegetables they wanted. Economic sociologists have shown that people often work for gifts and free items, particularly in markets where a concern for status is often more important than short-term earnings (McClain and Mears 2012; Mears 2011). In a similar way, the individuals filling the barn that morning saw social and moral value in the work they were doing beyond its mere economic rationality; they valued the goals of the farm and its CSA, and they valued the relationships and identities their work helped to create.

On a different Wednesday afternoon, after another hectic morning of filling CSA orders, Maggie and Joey sat down to weed a bed of carrots. The weeds were thick, and the session lasted for hours. They had gotten out of control, and there was the chance that three whole rows of red carrots could be lost. Their growth was already stunted when compared to the other rows, and they could hardly be seen beneath the much larger aggressive weeds. As the hours passed, we noticed the weather changing. The wind shifted, and the birds stopped singing. At that moment, John called on his cell phone to tell everyone to get out of the fields and come up to the barn. A customer had stopped by to pick up the week's CSA share and told them it was raining torrentially just up the road. In the distance we could see the approaching wall of near-black clouds.

It began to pour almost immediately. In light of the weather, the day's work ground to a halt. All of the workers went home, presenting us with an opportunity to talk to John and Katie while they waited in the barn for customers to pick up their orders for the day. We asked how long they had been farming organically at Scenic View. "Since the beginning," was John's reply. Of course, when he began farming ten years ago, there was no official definition of "organic," and there was no certification. He has only been certified for the past six years or so, since the USDA began regulating which farmers could legally call themselves organic.

John's decision to obtain organic certification was shaped in part by the opinion of his farming mentor—an individual whose critical influence on the farm will be given more attention in the next chapter. Often, bifurcation approaches maintain a skeptical stance toward certification, creating analytic categories pitting conventional-ized certified-organic farms against non-certified movement growers (cf. Constance, Choi, and Lyke-Ho-Gland 2008). It is often thought that larger-scale growers need certification for wholesaling opera-tions, whereas lifestyle and movement growers forgo certification because they sell directly to consumers or because the bureaucracy of certification makes it so tedious to maintain. However, for John and his mentor, getting certified was important for building an al-ternative farming movement. By getting official recognition from the USDA, countless farmers show the government and consumers that another way is possible in agriculture. If small farmers did not be-come organic and jump through the hoops, how was the govern-ment going to know that there were farmers who were committed to a sustainable vision of agriculture? People often expect imperfec-tion from NGOs that certify ethical production standards, and even with the standards themselves. However, perfect credibility is not necessary for such efforts to be viewed as valuable. Instead, the value of such efforts arises from the ability of individuals to fit the work of certification into a narrative they hold about how to make the world a better place (Gourevitch 2011). For Katie and John, getting certi-fied was first a way of understanding how their lifestyle, their every-day decisions, and their farming practices could be a part of a movement for large-scale agricultural change. Second, as John put it, "If you are going to put in all the extra effort of growing food or-ganically, you might as well get some sort of recognition."

The two seemingly unrelated interactions in the barn at Scenic View concerning CSA operations and organic certification reveal something fundamental about the social and economic decisions of Katie, John, and the other small farmers in our study. John and Katie's choice of the CSA as their primary sales strategy and their decision to keep their farm certified organic are both central to how they market their crops and run their business. However, both are shaped by and reflect the types of social relationships they have and aspire to create. As we came to learn, even the most seemingly "economic" decisions on the farm—like setting prices—were reflections of not just ideas about supply and demand or market conditions, but also of how John and Katie saw themselves as farmers and community members. These marketing strategies were also undertaken to foster the types of relationships that could fill the barn with the laughter of volunteers on a foggy summer morning. There are certainly places the types of relationships fostered by alternative agriculture could be deepened and made more meaningfully alternative and sustainable (see DeLind 1999). This is a problem we addressed in chapter 4 and one to which we will return in the conclusion, as we consider the future of sustainable agriculture. The examples here show the ways farmers' ideals and their customers' concerns are brought together in even the most presumably rational of marketing decisions—and in ways that the conventional agro-industry cannot replicate. Sometimes, they are even brought together in face-to-face relationships of gifting and support, lending the air of sociability that filled the barn that dewy Wednesday morning at Scenic View.

Supporting CSA

Organic farming is a lot of hard work. However, the CSA experience makes clear that there is much more to it than hard manual labor, a fact that only someone who has been around farms that were linked to surrounding communities would notice. John observed that Scenic View Farm was different from a lot of other small farms because of its CSA program.

"At this point in the season, other farms would be spending a lot of their time trying to sell their produce," John explained as the warm

July air blew in through the open windows of his minivan. "Because of the CSA, most of what we pick each week has already been bought." As a result, there were no frantic phone calls, no drives to distributors, and no heartbreaks on the auction block. Farmers like Lawrence Mendies (the New Jersey farmer who switched to a CSA after the agonizing decision to sell his zucchini for a paltry one dollar per bushel profit at auction) have learned the value of that assurance the hard way. John has likely never had to stomach the reality that the market value for his hard work was merely a dollar per bushel since he has utilized a CSA since the farm's inception. However, many farmers have had to do just this.

While the guaranteed closer-to-sustainable prices CSA subscriptions provide help farmers like Mendies, other farmers benefit from the way CSA programs help to stabilize the ups and downs of agriculture's unpredictable supply and demand cycle. Unlike other enterprises, a farmer can't necessarily ensure an even supply of crops across the growing season. Sometimes crops thrive, producing a surplus, and sometimes an epidemic—like the late tomato blight in 2009—wipes everything out. Even without the catastrophes that threaten crops each season—too much rain or too little rain, excessively cold weather or unseasonably high temperatures, diseases or pests—the seasonal nature of each crop variety makes it very difficult to ensure a steady supply. Diversity—in both the ecological and economic sense—on a small farm can help to ensure that some varieties will always be growing and that something will always be bought, but early spring will always be a time of low-profit spinach, baby greens, and not much else.

By receiving payments up front, farmers with CSA operations are less affected by the seasonal swings and, perhaps even more important, by the more unpredictable losses that could result from a bad year. In this way, "the CSA model of shared risk and rewards further collapses the differences in economic interests that normally pit consumers against farmers (a structural tension managed in the market through supply and demand price adjustments)" (Thompson and Coskuner-Balli 2007b, 296). Paul Atkinson, an egg farmer and owner of Oregon's Laughing Stock Farms, uses a CSA model to moderate the drop in income small egg farmers face in the winter months. While a chicken can lay an egg per day in the summer, the shorter days of winter suppress egg-laying behavior. Thus, while Atkinson's hens lay upward of four hundred eggs per day in the summer, he is lucky to get half that number in the winter. Such drastic declines in production could mean suffering through declines in income. Many farms avoid such declines by using artificial lights to induce winter egg laying, but Atkinson has chosen not to do that with his birds (Cagle 2011). The CSA model has helped make his decision to allow his hens to go through natural seasonal cycles economically feasible.

These types of benefits certainly helped Scenic View Farm. For example, Katie noted that the vegetables that the farm was able to provide in the early season CSA shares were kind of sparse. If it were not for the CSA, those sparse weeks would represent sparse income, precisely when expenses for supplies and labor on farms are highest. In the end, the bounty of summer and tomato season would represent a sudden windfall, when the farm's operating expenses had already been scraped together months before. For Scenic View, CSA shares were a balance of economics and ethics of sustainability and livelihood, a way to help make the farm profitable, as well as a means of supporting the types of agricultural practices John and Katie felt good about. Moreover, the CSA model matched the type of agrarian lifestyle John and Katie imagined for themselves and their family. It was the fact that everything that was picked on the farm was already purchased—except the limited quantities sold in the farm stand—that set apart a CSA farm in John's eyes. It allowed John and Katie to avoid scrambling quite so much. There simply wasn't the overwhelming pressure of produce waiting to be sold before it rotted.

For the other New England farmers to whom we spoke, CSAs were not a cure-all for the financial instability associated with farming's seasonal nature. As Sally said of the farm owner at Old Times Farm, "I know he struggles, financially, at times. Right now, things are good because there is money coming in. But in the winter, I know it's tough for him." However, CSAs served an important role in providing a level of financial stability and sustainability for farmers who had them. For Collin at Viridian, the CSA helped cover the season's operating costs. For Old Times Farm it gave a much-needed cash infusion after a difficult winter. And for Elaine at Pioneer Farm the CSA provided the type of guaranteed sales John and Katie thought were so important. To further level the financial instability wrought by the seasonality of farming, Collin at Viridian was introducing winter CSA shares, supplemented with products from other local food producers, and both Old Times Farm and Pioneer Farm were beginning to sell at indoor winter farmers' markets. Taken together, such efforts helped to provide the financial stability these farming operations needed to continue practicing the type of agriculture to which they were committed and to sustain the lifestyles to which their farmers aspired. In sum, they didn't need to hustle so much.

The CSA certainly seemed to make Scenic View Farm distinct from non-CSA farming operations. On top of providing some level of financial stability, it also had the potential to build community. The assembly line that assembled every week or so brought diverse people together around the food; the distribution of the food brought together another set of folks. Community can mean many things, but all agree that it means bringing people together to share in a common activity. The CSA was the kernel of such community building, and it operated outside of the mainstream channels of conventional food commerce. Not only was a significant amount of the harvest sold before the season even began because of the CSA, but also none of the produce ended up on supermarket shelves. Despite the grocery store being the primary context in which most of us think of food, all of Scenic View Farm's produce was directly marketed to consumers, either through the CSA, restaurant sales, or the farm stand. However, while John thought that the peace of mind provided by the CSA program was one of the greatest differences on his farm, it is just as significant to consider how Scenic View Farm was able to connect people to food through the CSA.

CSA customers are always at least nominally connected to their food supply, by knowing that all of their vegetables were grown on one farm by a specific group of people. The volunteer-based CSAs—involving people who take part in the production of their food system, helping to ensure that the farm's distribution goes smoothly—add a potentially deeply community-based component. With the Scenic View barn filled with people, alive with energy that Wednesday morning in June, the promise of organic agriculture embedded in community was palpable. Wrapped in the chill of a foggy morning, all of those people—in a barn of all places—were coming together around food for a common purpose.

The volunteers at Scenic View worked hard to balance the social and economic components of their relationship to the farm. They framed their work as voluntary, whether or not they were CSA members already. They were given gifts in kind, and workers and volunteers alike tried to make the packing of CSA orders feel more akin to helping out a neighbor than clocking in for work through their interactions and the social atmosphere they produced. "Perks" cement social relationships, and their gift-like qualities mean that they are not entirely interchangeable with cash: "The perk cannot be bought on any market" (McClain and Mears 2012, 140). Many of the volunteers were, in fact, paying CSA members who were packing their own bags of vegetables. However, the free items they received carried relational meanings they could not get with a purchase from a farmers' market or even with a CSA share. Katie stressed that she wanted the volunteers to be "taken care of" when she described the free vegetables she sent home with them: they were not paid like employees but provided for as friends. These social negotiations were good matches in that they managed to get the economic work of running a CSA accomplished without blurring the lines of friendship and community. Paying attention to such negotiations is critical for understanding not only the contemporary organic sector beyond its structural political economy, but also the types of economic and social-relational work that could lead to a more truly alternative and sustainable agricultural system.

What brought together such a diverse and interesting group of people on a foggy Wednesday morning to load vegetables into a sea of brown bags? For some, it seemed to be a passion for an agrarian ideal: working together in and with the elements. The common goal was to produce (and eat) high-quality healthful vegetables, free from pesticides and synthetic fertilizers—that is, organic vegetables. For others, the purpose seemed to be a desire to support those engaged in such efforts. Their practices and relationships were the product of constantly negotiating and balancing economic, social, and personal commitments—creating an economy where labor was gifted in exchange for the satisfaction of one's having played a role in a more sustainable agricultural system. For some, it was a means of self-sufficiency: a morning's work meant a degree of agency in the food system. For others still, the reward was the chance to get out of the house, connect to new people, and leave with a free bag of healthful, organic produce.

Each individual's level of commitment to organic agriculture varied, as did the reasons for that commitment. Nevertheless, the concept of "organic" remained central to the farm's collective identity and served as a basis for connecting these people from the surrounding area to the food they ate.

Contesting Organic Certification on the Ground

Other small-scale farmers with whom we spoke, at certified organic farms like Parnassus Farm and Laughing Brook Farm, had similar feelings about the importance of organic, both as a principle and a marketing strategy. As Meghan of Sustainable Harvest farm mentioned, the head farmer under whom she works strongly believes in the importance of certification as a matter of principle, much like John and Katie. Likewise, Elaine at Pioneer Farm saw the value of certification, first because it was important to many of Pioneer Farm's customers and second because it was a way of signaling a baseline of sustainable practices to consumers who might not get to ask the farmer—or feel comfortable asking—about how their food was grown: "People look for [certification]. It's good for consumers who don't get to talk to the farmer directly or go to the farm and know for sure." Molly at Grateful Harvest Farm felt similarly. Her farm began the process of organic certification because its customers cared about it. But none of the farm's practices changed as a result of this transition. Even though her farm has chosen to certify, Molly suggested, "Knowing how *local* and how *ecologically healthy* the practices of the farm's product are has more weight on my purchasing decision than reading the words 'certified organic.'" Nevertheless, for her small farm organic certification has become an important part of its marketing strategy to secure consumers.

Even Collin, who decided not to continue certifying the produce on Viridian Farms after the USDA took over the certification process—because he saw it as no longer being as meaningful—saw the value of certification. "I think certified organic is really important for consumers. It lets them trust that the word 'organic' actually means *something*," Collin explained. "But given the fact that we market direct to consumers, we just felt that our customers were ready to trust us as much as they trust the certification." Collin added, "I don't object to the way certification is currently being done." However, Collin responded to the issues surrounding certification differently than did the six farmers growing USDA organic produce at Heritage Harvest, Parnassus Farm, Pioneer Farm, Laughing Brook Farm, Sustainable Harvest, and Scenic View.

Endorsements of the USDA's certification program can be difficult to find above the complaints from small farmers and popular accounts of the organic farming movement. And those complaints abound. Apart from arguing that the legislative process has watered down organic standards—an argument that has been successfully levied by both academics (see Guthman 2004a, 2004b, 2004c) and small farmers—many small-farm operators claim that the whole process makes onerous demands on their time and finances. One of the most prominent critics is Joel Salatin, owner of Virginia's Polyface Farm, who features prominently in Michael Pollan's *The Omnivore's Dilemma* (2006). His rhetoric is captivating. For example, Salatin had the following to say regarding organic certification: "You know what the best kind of organic certification would be? Make an unannounced visit to a farm and take a good look at the farmer's bookshelf. Because what you're feeding your emotions and thoughts is what this is really all about. The way I produce a chicken is an extension of my worldview. You can learn more about that by seeing what's sitting on my bookshelf than having me fill out a whole bunch of forms" (cited in Pollan 2006, 131–132). He also quipped: "We never called ourselves organic—we call ourselves 'beyond organic.' Why dumb ourselves down to a lesser level than we are?" (132). Such interactions with Salatin on his farm led Pollan to conclude, "There is this further paradox: Polyface Farm is technically *not* an organic farm, though by any standard it is more 'sustainable' than virtually any organic farm" (131).

Other commentators on small farms and in organic agriculture echo the same refrain. For example, Collin Beavan recounts asking the leaders of a sustainable food organization whether the local farmers he was likely to encounter at farmers' markets in his *No Impact Man* experiment would be organic. He was told, "Many of the farmers [he'd] find in the farmers' markets considered themselves 'more organic than organic'" and that "Most of them are part of the old-fashioned organic movement" (2009, 125). Upon asking how he could tell, without certification, if they used organic practices, they responded, "You look them in the eyes and you ask them. This is the thing about local food. . . . Ask them what their priorities are, and if those priorities line up with yours, they're the ones you buy from" (126).

Similar statements rejecting the official certification regime seem endless. For example, Mark Rhine, a farmer from Phoenix, reflected, "When you get through the application, and you realize how much it would cost us to grow the way they wanted just to use a label, we thought it wasn't worth it" (cited in Silverman 2011). Vicki Westerhoff, who utilized organic practices for eleven years before getting certified in 2010, remarked, "For me, it's a lot of work and I don't think organic certification is set up for the small farmers anymore" (cited in ibid.). She chose to get certified only because the farmers' markets she frequents began requiring vendor certification. A common sentiment among small farmers seems to be that with local agriculture, organic certification is meaningless. One such farmer is Henry Brockman, who felt there were few cases where certification was important: "The only way I see certification being necessary is doing some bulk wholesale to Whole Foods" (cited in ibid.). Local farmers can look a questioner in the eyes and say they are sustainable, developing a reciprocal relationship of trust. Better yet, many local farmers encourage patrons to visit their farms to see the farms' practices for themselves.

Many of the New England farmers in our study expressed similar sentiments. Collin at Viridian made the decision not to be certified because of his farm's direct marketing approach—he felt that certification was valuable but not meaningful when customers could speak to him directly about what he was doing on his farm and could trust him as much as they can trust a label. Sally of Old Times Farm echoed the concern that certification wasn't set up for small farms any more. Old Times Farm has a rather small, twenty-two-member CSA. As Sally explained, the costs of certification were too high for her farm to bear, and the added labor of record keeping was too much for a farm with merely two permanent employees. The practices on Old Times Farm, however, have not changed as a result of the decision not to maintain organic certification after the USDA began regulating the process. For her, Old Times Farm's produce was still "organic." Sally just cannot call it organic at the markets at which she works.

The problem with the "beyond organic" discourse surrounding organic certification is that it fails to capture the whole story. For one thing, for the majority of our non-certified but sustainable farmers in New England, the choice to forgo certification was not because of a desire to be "beyond organic." Rather, it was a practical decision that the strictly bureaucratic regulations of certification were incompatible with the real world of agrarian life—a life where a year's worth of literal blood, sweat, and tears can be undone in nearly an instant by unforeseeable pestilence, disease, or weather.

As we highlighted when discussing the outbreak of late tomato blight on farms across New England during the early summer of 2009, many farmers forgo putting the time, money, and added effort into organic for fear that they will only need to spray to prevent the total loss of a crop. By forgoing certification these farmers hardly see what they are doing as "selling out" or "cutting corners." These farmers are making conscientious decisions—like Jane and her colleagues at True Friends Farm, who stressed that when the farm's Integrated Pest Management program recommended spraying, they tried to spray as long before harvest as possible and use only the safest sprays available. Farmers like Eric at Johnson Family Farm or Matthew at Peaceful Valley Orchards talk about being "responsible" in their farming practices and "wanting to do the right thing." As Matthew said of his farming practices, he wants to be able to look his customers in the eye and be able to tell them he feels comfortable with the way he is farming.

Perhaps, above all, these are farmers doing their best with what they know—both about agriculture and about themselves. As Michael Bell explains, "It is through knowledge that farmers farm the self" (2004, 129). By trying to be good farmers—the best they know how—these individuals work to construct an image of themselves as good people. For these farmers, the decision not to get certified was not the result of their feeling that their practices surpassed the USDA organic regulations—that they were "beyond organic." These were practical decisions made by individuals trying to make their ethics work in contemporary agriculture. For them, the bureaucratic USDA certification process could not accommodate the uncertainty of farm life. And while they did not end up certifying their farms like the farmers at Scenic View, their concerns about being responsible farmers were no less real.

Such "beyond organic" discussions have led consumers, as well as commentators like Michael Pollan, to conclude that non-certified farmers using organic practices—like Joel Salatin—are somehow "more 'sustainable' than virtually any organic farm" (Pollan 2006, 131). In some cases, that may be true if by "virtually any organic farm" Pollan means virtually any industrial-scale, supermarket organic farm that we discussed in the beginning of this book. Moreover, it has led many farmers like Salatin to revile the very word "organic"

Figure 6. Managing Pests: Orchard crops, like these apples, were often cited as making formal organic certification impractical. (Photo courtesy of Connor Fitzmaurice.)

as a dumbed-down facsimile of true sustainability. Yet Salatin is now something of a celebrity, giving paid campus lectures, and he has a large production farm. His is a deeply privileged position in which to be and from which to reject the benefits of the USDA organic label. Smaller farming operations, such as Scenic View, cannot make rejections so lightly. For certified small farms like Scenic View, Pioneer Farm, and Heritage Harvest, "organic" is still a word that has meaning to them personally and to their customers, and is seen as providing them a level of security by generating consumer confidence in their production practices.

When Organic Certification Makes Sense Locally

As the case of Scenic View Farm demonstrates, there are small-scale farmers who are doing organic well—even within the USDA regulatory framework. Organic agriculture and the local and sustainable food movements are not necessarily mutually exclusive for every small farmer. There are certified-organic small-scale alternatives to the corporate organic products available at Walmart Supercenters and on Whole Foods' shelves. Despite the rhetoric, some small sustainable farmers willingly *choose* to become certified organic. And these farmers need to be accounted for in the broader discussion of U.S. sustainable agriculture.

John and Katie's farm is one of them. And it would be hard to imagine that they haven't heard the raucous protests against certification from farmers like Salatin. (As mentioned above, Michael Pollan's books were virtually required reading for Scenic View employees.) Likewise, John's certification was not a desire to "go big" and sell wholesale to stores like Whole Foods. As a direct marketer, he could have forgone certification and told each of his customers how he farmed, and they would undoubtedly have believed him. Most of all, John and Katie were not duped and blindly following the regulations. Yet they willingly chose to jump through the government's regulatory hoops, pay for third-party inspections, and become certified.

For some small-scale alternative farmers organic certification no longer makes sense. However, theories of bifurcation, which seek to deal with the polarizing debate about the regulation and cooptation of organic, fail to recognize the nuanced ways small farmers respond to changes on the ground—changes that may not line up with binaries of "market" versus "movement" or "certified" versus "beyond organic" growers. Certification still matches with the lifestyles and realities of certain small farms. In order for us to explain this asser-tion, however, an ethnographic approach attentive to the negotiation of such matches is necessary. The bifurcation of the organic sector is not so cut and dried. Instead, farmers come to understand their decisions regarding certification in ways that balance the economic costs and benefits of certification with the types of relationships they hope to cultivate in their fields and communities—and with their vision of themselves as farmers.

Many small farmers at farmers' markets talk about how they opted out of certification because they too were "beyond organic." But John and Katie believed that becoming certified was the only way to let the USDA and the consuming public know that there were still small farms involved in organic agriculture. For Scenic View Farm, becoming certified wasn't settling for—or, more cynically, cashing in on—a corrupted and watered-down concept. It was more like casting a vote for a new vision of America's agricultural system. Or, in John's exact words, it was "like joining a movement." This is not to say that John believed that the current state of affairs for organic agriculture was perfect or that the critiques of farmers like Salatin were not valid. However, John seemed to feel that in order to change the agricultural system in this country—and not just carve out a niche within it—farmers needed to work within the USDA frame-work and demonstrate that there was a need and desire among farm-ers to be sustainable. How else would federal resources be mobilized to make sustainability a reality in America's agricultural economy?

Beyond such issues, the polarizing debate about local farmers being "beyond organic" neglects important practical issues that the organic certification system—despite its many flaws—was enacted to prevent. For example, even if consumers can simply ask a local farmer—or visit a local farm—to learn about farming practices, they may not be in a position to determine whether those practices even meet organic standards, let alone surpass them. The average organic consumers—confronted with a beautiful organic farm like Scenic View—might not think to ask if the farm uses cover crops to pre-vent erosion in the off-season, rotates crops to prevent disease and nutrient depletion, or gives its fields a fallow year. They may simply see a diversified farm, ask about pesticides and fertilizers, and be satisfied. The fact is that the basic questions that most of us as consumers are knowledgeable enough to ask cannot possibly reveal whether or not a farm is truly organic—let alone "beyond organic."

And there may be a more important reason not to give up on the organic label just yet. In 2010, NBC Los Angeles conducted an undercover investigation at several farmers' markets in the Los Angeles area. In one case, the investigators followed the traditional advice and "looked the farmer (Juan Uriostegi) in the eyes," asking if everything that he was selling was grown on his farm in Redlands, San Bernardino County. Uriostegi assured the investigators it was. Having bought some broccoli, the investigators then visited the farm with a San Bernardino Department of Agriculture officer. When asked where the broccoli was grown, Uriostegi could only point to a dry patch of dirt (Grover and Goldberg 2010). More alarming, the investigators purchased five containers of strawberries from five different vendors, all containers advertised as "pesticide free." After they were tested at a state-certified lab, three of the five containers were found to have levels of pesticide that were higher than would be expected from accidental contamination (ibid.).

Certification can also protect the identities of alternative small farmers in ways simple farmers' market sales cannot. Many farmers' markets include wholesale produce that a "farmer" buys and then sells at his or her own stand. Only at "producer farmers' markets" do the people who actually grow the food sell that food. Many customers that frequent farmers' markets do not know this. As one professor of sustainable agriculture in Illinois told us, "I've actually seen where Amish farmers buy tomatoes and peppers from a wholesaler and sell them at the local farmers' market. Consumers automatically think the Amish farmer produces 'better' veggies—and he hasn't actually produced anything" (personal communication).

Our own conversations with New England farmers revealed some of the same risks with "simply looking the farmer in the eyes and asking." When we began to talk to Sean, of Verdant Acres Orchards, he immediately said that all of the produce his orchard sold was organic. "But sometimes we have to spray things to fight disease so we can have all the beautiful produce you see here." When pressed about how the farm was organic but still spraying, he responded, "Oh, we're certified." Assuming that he was then describing sprays included for use on the USDA's National List of Allowed and Prohibited Substances, we asked what it meant to him, personally, for the farm to be organic. "Oh, well, we're not organic," he responded. "The way I see it is, when people come up to you and say, 'Are you an organic farm?' it's just a farmers' market saying. That's all it is." Lost in the "beyond organic" argument is the fact that not all local farms are as committed to practicing alternative agriculture as are the people at Scenic View Farm and many of the other farmers in our study. For some, "organic" is just a farmers' market saying designed to entice urban consumers.

This is hardly conclusive evidence that farmers' markets are plagued with fraud. However, despite the arguments that organic certification is not important when consumers are able to meet their local farmers, consumers are simply not in a position to determine just how organic a "beyond organic" farm is. Certainly, many of these farmers are more sustainable than the organic regulations require, but the average consumer lacks the resources to make such a determination. For farmers like John, who are genuinely committed to sustainability and whose direct-marketing practices could obviate the need for certification, the record keeping and inspection requirements of certification help keep them even more accountable to their customers. They feel that they can't slip below a minimal standard of sustainability because of this transparency. Yet other farmers avoid certification for the same reasons. As the tomato blight in 2009 demonstrated, organic standards don't always match well with the local ecological conditions of production that small farmers come up against. You can lose it all, or you can lose the label.

In addition to transparency, John and Katie feel that the meticulous record keeping required for certification, about which so many small farmers complain, is a type of "best practice" for organic farms.

Because organic farmers cannot rely on chemical quick fixes, having detailed records of what was done in the past helps them to plan better management strategies. For example, it helps them to know which intercropping methods worked before and which fields were not producing well. In effect, they feel that their decision to become certified—and the structural requirements that certification imposes— helps them to be better farmers. Every time a seed is planted, a crop harvested, organic fertilizer applied, or an approved method of pest-control employed, John and Katie have to make a record of it. These records are then submitted to their certifier, in addition to annual inspections.

Moreover, after we observed and participated in life on the farm, it was easy to see why any recognition—even if it were from a government with watered-down standards—would be welcome at Scenic View. Just as CSA Wednesdays illustrated the promise of organic farming when embedded in community, life on the farm also bore witness to the challenges small farmers face when attempting to work out their agrarian ideals and ethics in practice. Growing vegetables organically was a constant struggle. Every day there was some problem to be addressed, from the moment work began to the moment we stopped for the night. Katie put it best one afternoon: "Every year there are problems, and every year they are different."

Putting Organic Values to the Test

In 2009, with the onset of the late blight afflicting northeastern tomatoes, John and Katie weren't exactly nervous about the tomatoes. What made them nervous was what they discovered when harvesting the first new potatoes of the season. Moving along on hand and knees, uncovering the small potatoes from beneath the farm's rich soil, John became concerned. He noticed several seemingly diseased potatoes.

"This tomato disease affects potatoes too," he said, obviously frustrated. "It's the same disease that brought the Irish to this country." John had everyone walk down the rows of potatoes and inspect each plant for blighted leaves. Donna called out that she found one; Joey found another. John dug up several potatoes from beneath these plants, checking for blight. It had not made it to the tubers on all of the plants he checked. Perhaps the initial finds were an anomaly. John would have to check again, and if the blight showed up, all the potatoes would be pulled prematurely.

Beyond provoking fears of blight, the 2009 potato crop at Scenic View provided another example of the types of problems faced by organic farms year after year. Every row of plants had bloomed, signaling that the potatoes beneath the ground were forming. When the plants died back, it would be time to harvest the crop. For now, however, the field was a sea of dark green, punctuated by beautiful clusters of lavender flowers. (Commenting on their beauty, Joey remarked that Marie Antoinette, always one to [supposedly] dictate the eating habits of her subjects ["*Qu'ils mangent de la brioche!*"], took up wearing potato flowers in her hair to help popularize the then strange New World vegetable.) But the field only seemed beautiful from afar. John and Katie, who had been monitoring the crop more closely, knew that every plant was at risk not only for blight; the plants were already covered in potato beetle larvae that were slowly devouring the visible plants, potentially stunting the new tubers growing just beneath the surface of the soil.

One afternoon, Katie reported that the potato beetle larvae on the plants were maturing, and she hoped to spray the plants with a bacterial preparation to combat them before they metamorphosed, mated, and began the cycle of destruction anew. The farm had begun using an organic spray the year before, with some success. The bacteria in the spray get ingested by the bugs as they eat the plants and then "eat apart their stomachs," killing them. The problem, however, is that once the plants flower, bees are attracted to pollinate the plants. As they ingest nectar from the lavender blooms, they too ingest the bacteria and are killed. As a result, the potato blossoms could not have come at a worse time.

The solution was a simple yet labor-intensive one, reflecting the deeply held values guiding the farm. John instructed everyone to cut off every single potato flower so that Katie could spray the plants; they were going to have to "waste" valuable working hours in the middle of the growing season to protect the already vulnerable bee populations, especially wild bees (cf. Pisani Gareau et al. 2009). As we approached the plants, sheers in hand, the beautiful scene transformed into a nightmare. The dark green "leaves" seen from across the farm were mere skeletons. Writhing, terracotta-colored larvae had covered the plants by the thousands. Their soft, black-spotted bodies undulated rhythmically with the endless motion of their mouths. All that was left of the leaves were the tough central veins, and even these were being eaten. John could have just sprayed the plants; no one would have known. But down each row we went, sending lavender blooms fluttering softly to the ground by the thousands.

A farm uncritically applying bacterial preparations to combat a potato beetle outbreak may represent a conventionalized practice of mere input substitution. However, in the fields of farms like Scenic View, farmers carefully negotiate the practices with which they are comfortable and under what conditions. John and Katie feared losing potato plants to a particularly bad outbreak of potato beetles. They knew there were allowable organic sprays available that could help prevent these losses; they also knew that the presence of flowers on their potato plants meant that the use of these allowable sprays could endanger pollinators like honeybees. So John and Katie made the best choice they perceived as being available to them: they used the spray, but only after removing every flower from the plants. Farmers like John and Katie demonstrate that behind every organic practice are pragmatic decisions that seek to balance concerns about the economic sustainability of their farms as businesses with equally serious concerns about the sustainability of their farms as members of social and ecological communities.

Struggles with potato blight and potato beetles are merely examples of the daily trials at Scenic View Farm; there were countless more. Aphids devoured the chervil, hornworms infested the tomatoes, and cucumber beetles scoured both cucumbers and squash. Mildew ruined the last of the peas, mice ate the beets, and leaf miners left the arugula marred with labyrinthine white marks. In the end, each of these problems could have had an easy chemical solution. The tomato blight could have been forestalled, if not prevented, by fungicide. The potato beetles could have been wiped out with a more powerful pesticide. And were the farm not organic, no one would have cared if a couple of hundred, or even a thousand, bees died along with them. In light of such easy solutions, why be organic?

Standing in the barn one Wednesday afternoon, the rain beating down on the metal roof, John and Katie mused about the costs of their organic commitment. Did they ever think about how much money they would make if they didn't lose so much of their crop to diseases and pests?

"To tell you the truth," John began, "I have never even thought about it that way. I'm not going to be conventional no matter what, so I never considered those sorts of savings an option."

Katie also admitted to having never thought about the monetary losses they faced due to being organic. "With how much I worry as it is," she continued, "I can't even imagine how I would feel if I was constantly thinking that all of my problems would be solved if I just sprayed the plants with poison." Such losses are not unique to Scenic View Farm. "Every year there are problems" could be a slogan for the organic farming community. Certainly all farmers—not just organic and sustainable farmers—face problems every year, and it is not as if Katie and John's financial problems would be over if they simply sprayed their crops with pesticides and fungicides. Describing the challenges of contemporary agriculture—both conventional and sustainable—Bell rightly notes, "The uncertain landscape of farming is something with which [farmers] all must contend" (2004, 160). However, the decision of farmers like Katie and John to maintain organic certification complicates the issues they face from year to year. But while all farmers may face the vagaries of both pestilence and weather, sustainable farmers have even more to fear; they are unprotected from such misfortunes by agricultural subsidies and less able to keep lenders committed to their viability since they spend and borrow less on inputs (237). The result is that "sustainable farmers face several disadvantages in the farming rodeo" (ibid.). And while Katie and John have never contemplated the economic burden such problems place on their farm, other farmers have.

Many New England farmers who chose not to certify their farms certainly have. Especially for farmers choosing to employ an Integrated Pest Management program or sustainable no-spray/low-spray practices rather than certified-organic practices—like those at Peaceful Valley Orchards, Long Days Farm, and Johnson Family Farm—their decisions have been motivated by the knowledge that to avoid the total loss of certain vulnerable crops to disease, they may occasionally need to reach for the fungicide. These farmers strike a different balance among economic concerns, ethical commitments, and lifestyle considerations in their farming practices. John and Katie have focused on the added value that organic provides for their produce, the ecology of their landscape, and their perception that being organic allows them to most fully enjoy that landscape—even through the simple pleasures of eating a carrot straight from the ground. All of these farmers, however, seek a balance among certification, lifestyle, and their bottom line—both fiscal and ethical. All of them want to be responsible farmers—a testament to the ability of alternative farmers to remain true to a vision of "good farming" in the face of tremendous economic risks.

Often, farmers must navigate tremendous possible risks when deciding to become certified organic. Take, for example, New Jersey's blueberry industry. One of the nation's top producers of the fruit, New Jersey had three hundred farms growing blueberries as their primary crop in 2011. Only four of those were certified organic (La Gorce 2011). Such scarcity of organic growers is not due to a lack of demand. John Marchese, who has twenty acres of organic blueberries, says he can never seem to grow enough to meet the demand. "Every year we get bombed—the phone never stops ringing and people never stop coming. They ask for directions from the Holland Tunnel, or as they're crossing the Delaware River. It can be completely overwhelming," Marchese admits (cited in ibid.).

In large part, the scarcity of organic growers has to do with the losses they can expect. According to Peter Oudemans, a Rutgers University professor and plant pathologist, organic blueberry farmers can expect up to 50 percent of their crop to be lost to disease or pests, compared to the 5–6 percent losses typical among conventional growers (La Gorce 2011). Dr. Oudemans described the situation with a colorful analogy: "Planting a solid acre of organic blueberries in New Jersey is like throwing a peanut butter sandwich into a room full of kindergartners" (ibid.). Blueberries, as a native crop in New Jersey, face a predator base perfectly adapted to eating them.

It's not just that farmers who are looking to grow challenging crops in certain environments must weigh the costs of farming organically. For struggling small farms, it is simply not economically feasible to lose an entire crop. For Kira Kinney of Evolutionary Organics in New Paltz, New York, the tomato crop bankrolls her business. " 'Tomatoes get me out of debt every year,' " Kinney said. " 'I go into the season with credit card debt and I come out O.K. That's how I cover my annual costs for the whole farm' " (cited in Moskin 2009). John and Katie were in a similar position at Scenic View. John told us that the tomato crop was the farm's biggest source of income, making the 2009 tomato blight a particular cause for concern. He was worried, but for some reason he said he never thought about such losses as a consequence of being organic. Based on the security he felt he had thanks to other organic practices—like a diversity of crops and a CSA model that shifted some of the risks from his certification onto consumers—organic certification remained a viable match with the local conditions of his farm. Without the CSA, organic certification might not have been such a seemingly commonsense marketing strategy for the folks at Scenic View.

Why do some organic farmers seldom think about their situations in this light? Any of the near disasters described above would have most people thinking about the alternatives. The thousands of writhing potato beetle larvae covering the potato plants can be a morale-draining sight. For some alternative farmers, it is. They choose to work out their visions of what farming responsibly means without subscribing to the narrow, input-avoidance model of large-scale organic agriculture out of concerns that their small farms potentially couldn't handle a serious loss of a crop to disease. However, for others—like those at Scenic View—local conditions allow organic certification to be viewed as a viable match between the types of agrarian relationships and lifestyles they hope to cultivate and their economic constraints.

For John and Katie, there was only one way they were willing to farm—organically. As a result, such struggles just came part and parcel with farming, not a particular type of farming.

Such commitments are echoed in the response from Sally (of Old Times Farm) when she reflected on whether her farm's practices would ever change: "Oh no. No, no, no, no." They are reflected in the assertion of Collin (of Viridian Farm) that his farm's practices are just as organic now with a fifty-acre farm as they were when he started with a ten-acre farm; they just involve more workers and more equipment but no less concern for being sustainable. And they are mirrored in the earnest desire of farmers like Matthew of Peaceful Valley Orchards who wanted to be "responsible" and do right by his farm and consumers by striking the right balance with their farming practices. Matthew spoke with considerable pride when he shared that his farm hasn't needed to bring in bees to pollinate the fields and orchards in almost twelve years "because the way we are working and the way we are farming has allowed the natural population of pollinators on the farm to thrive."

This is not to say that farmers like John and Katie at Scenic View would not be farming if it were not for the USDA organic certification system established under the National Organic Program; after all, Scenic View was farming "organically" before there were federal regulations governing organic foods. Moreover, they could easily fit into the "beyond organic" crowd. In their farming practices, the workers at Scenic View sought to uphold the ideals of the organic movement and not just the letter of the law. They would not use the exception for copper sulfate to solve the problem of blight, and they would not irresponsibly apply biological controls that could harm pollinating insects just because it was allowed. In many ways, the practices at Scenic View suggested that its workforce also felt that the organic regulations were watered down. Nevertheless, John and Katie chose to certify the farm.

In becoming certified, John and Katie were not endorsing a watered-down vision of organic farming practices. Rather, they pragmatically felt that if the label was there, it should be used. In a way, it makes sense. Didn't a majority of the organic movement advocate for recognition for decades? Movement actors were behind the push for federal recognition of organic for many of the same reasons that John and Katie still choose to be certified: it could ensure consistent meaning of the term for consumers and potentially lead to greater USDA support and research. And unlike some approaches to the organic market's bifurcation might suggest, certification is not synonymous with conventionalization (Constance, Choi, and Lyke-Ho-Gland 2008). For some small farmers, the practices mandated by the National Organic Standards Board are still a good match with their lives and their concerns for their livelihoods. If there seems to be a contradiction between Scenic View's "beyond organic" practices and its participation in the USDA organic program, it is only because our current agro-economic system forces many such contradictions upon farmers who attempt to farm in a manner that recognizes sources of value beyond economic utility. As we discussed above, the history of organic is messy, and the meaning of the word "organic" may be losing value with some consumers (see Adams and Salois 2010). However, farmers like John and Katie must work within a system that forces sustainable farmers to deal with countless contradictions, carving out room for sustainability in an agricultural system that is often utterly unsustainable—regardless of whether they choose to become certified.

Prices as Social Relations

"John," Katie called as she entered the field one afternoon. "The chervil is going to seed already."

"I guess we'll have to try pinching it back."

"I did," Katie replied. "And it has aphids."

"Well, hopefully the ladybugs come soon."

There was literally nothing to be done. John, out of his commitment to organic agriculture, had resigned himself to the forces of nature. If nature dealt a blow in the form of aphids, nature would have to provide its own solution. Within a few weeks, nearly every chervil plant was dead or stunted.

As the weeks went on, numerous similar events would demonstrate the delicate ways in which market forces were balanced with ethical and lifestyle decisions through the farm's organic practices. Often, there seemed as though there were far too many difficulties to contend with: blight, potato beetles, cucumber beetles, fungi, aphids, cutworms, and leaf miners. And each of these struggles had to be matched with practices with which both John and Katie felt comfortable and that made sound business sense for the future of Scenic View.

John and Katie were clearly not getting rich off of farming. Rather, it was John's aspiration for his family to someday enter the ranks of America's middle class. Moreover, the very premise that market forces motivate organic farmers is based on the assumption that they all receive a premium for organic crops. This assumption seems true when one is walking through most supermarkets that sell organic produce: the organic costs more. However, over the course of our participation at Scenic View Farm, it became clear that organic agriculture is not always about profiteering. Although Katie had once mentioned that some of the losses they incurred by farming organically could be recouped by the ability to charge more, the markup they received was never obvious. "Do you know what the prices are like at the city markets this season?" Katie asked one afternoon as we stood in the barn's farm stand admiring the bounty on display. That week, no item was priced higher than $2.50 per bunch.

"Much higher," Connor Fitzmaurice said, pointing to a bunch of baby turnips. "One farm was selling those for $4.00 a bunch."

Katie was shocked. Her turnips were being sold for only $2.00 a bunch. She walked over to a basket in the corner and pulled out a piece of chalk. Smudging the zeros with the side of her hand, she raised the price a mere $0.50. "I hate not giving people a good deal," she said, "but we do need to make money." Still, $2.50 was hardly expensive.

The seemingly natural "bunch of greens" reflects a conscious decision by the farmer as to how much an average customer wants, how much the greens are worth, and what the farmer thinks is a marketable unit price. In other words, there is nothing "natural" about a bunch of greens. They certainly don't grow in even-sized bunches. However, prices are also a means of communicating social values—not just economic ones. In art markets, for example, rising prices are a sign of an ascendant new talent. Conversely, gallery dealers rarely lower prices on works of art that don't sell because this would signal a poor evaluation of the artist's reputation and talent and not just a miscalculation of what the work of art was worth (Velthius 2004). In the same way, putting together a larger than average bunch of greens for a reasonable price was a way for these farmers to try to signal their concern for their customers and send a message about the type of business they were running. As a result, the seemingly mundane process that Katie and John went through to decide how large the bunches of each crop should be often revealed the subtle economic and social calculations that shaped the farm's practical decisions. After showing an example of an appropriate bunch, Katie would invariably seem to add an extra stalk, saying, "You want to give people a good deal." Giving consumers these sorts of "good deals" matched the needs of the farm in more ways then one. In fact, Thompson and Coskuner-Balli argue, "In symbolic affirmation of nature's bounty, CSA boxes frequently contain more of a given vegetable (or vegetables) than consumers find that they consume before spoilage sets in. This irrational volume of goods—in this sense of not being optimally synchronized with household consumption rates—precipitates sharing networks and, hence, feelings of communal connection" (2007b, 287). Like the running of a CSA or the maintenance of organic certification, marketing practices surrounding pricing were reflections of the farm's real and aspirational social relations. To be reasonable, prices had to viably match the sense of fairness that John and Katie shared, the vision of a food system they hoped for and that ensured the types of satisfied customers who would buy another share—and might even come to the farm to help out with the formidable task of bagging food for one hundred families.

Efforts to give customers "good deals" shaped everything from the size of a bunch of arugula to the price of turnips in the Scenic View farm stand. These concerns went straight to the top of the farm's handling of revenue—even shaping the pricing for CSA shares. Katie explained that the farm hadn't raised the price of shares in its CSA for years. She felt that they needed to—and she and John had even decided that they were going to—but then the economy tanked. Katie said that they couldn't raise the prices then if they wanted their food to remain affordable. Maybe when the economy turned around they would, but in 2009, a year after the recession began, raising prices remained off the table.

Many people might expect organic farmers to be conscious of cutting their costs everywhere possible and charging premium prices; they would not expect that the seemingly mundane decisions of such farmers, like rubber banding bunches of kale, would be governed by their desire to give people a bargain. Such negotiations mean that John and Katie arrive at prices for their produce with which they feel comfortable, even while sometimes failing to capture much of the price premium associated with the organic label. Such prices weren't just poor marketing decisions on Scenic View Farm; they reflected a desire to keep their produce affordable and an ethical commitment to make it more widely available to consumers in their community, a commitment shared by many of the other farmers in our study.

Although Elaine of Pioneer Farm felt that her prices were higher than what one would pay for conventional produce in the supermarket, she was deeply concerned about making organic local food more accessible. "That's something that I hope we, not just as a farm but as a community, can figure out: how to get the prices of organic food more affordable for the majority of consumers." Both Mary and Jane of True Friends Farm talked about the importance of people's having access to fresh, healthful food—a position that their farm has addressed by choosing to accept WIC payments (Women, Infants, and Children assistance from the USDA's Food and Nutrition Service). "Our farm is located in an area where a lot of people are on WIC," Jane reflected, "and it's important that our community [not just the urban markets to which True Friends sells] have access to healthy food." John and Katie and the other New England farmers weren't being poor business people by "trying to give people a good deal." Across many of the farms where we interviewed, such decisions reflected commitments to make organic food accessible as something more than the "yuppie chow" some academics have characterized it (Guthman 2003).

John and Katie's decision not to raise prices in 2008, and again in 2009, seemed to be at odds with broader trends in the organic food sector. Leading up to the recession, organic food prices were increasing at a slower rate than conventional prices due to an oversupply of organic milk, the introduction of private-label organic products by large retailers, and a general reticence to raise already high prices (Martin and Severson 2008). However, by 2008, prices had skyrocketed. Gary Hirschberg, CEO of Stonyfield Farm, described it as "probably the most dynamic and volatile time I've seen in 25 years" (cited in ibid.). Organic Valley milk increased in price from $3.49 to $4.00 for a half gallon. In 2007, Eggland's Best organic eggs sold for between $3.79 and $4.29 per dozen; in 2008, they were selling for between $4.59 and $4.99 (ibid.). As the prices of organic grain rose, some dairy farmers were forced to abandon organic production altogether, while others complained that retailers were not charging enough for their products (ibid.).

Nationwide, the premium charged for organic fresh produce displays considerable variability. Studies have shown premiums above conventional prices of 15 percent for carrots to 60 percent for potatoes, and on average, U.S. consumers pay an additional eighty-six cents per pound for organic strawberries (Lin, Smith, and Huang

2008). With such extreme variability among merely fresh produce, the premium for all categories of organic foods in the United States is estimated to range from 10 to 100 percent (Lockie et al. 2006, 107). These types of premiums were what we anticipated when we went to Scenic View Farm. However, much like the "beyond organic" rhetoric, the public discussion surrounding organic food prices reveals only part of the story.

Indeed, despite the popular belief that local organic food costs more, the data do not consistently bear it out. For example, in Iowa, a study compared food prices from the farmers' markets to those of conventional grocery stores (Pirog and McCann 2009). The study used a market-basket approach, looking at a hypothetical shopping order consisting of popular seasonal vegetables. Contrary to expectations, the average price per pound of the local farmers' market basket was $1.25, while the average price of the non-local basket at the supermarket was $1.39. "If a family of four was to purchase half the Iowa per capita consumption of each vegetable in the vegetable basket . . . the local vegetable basket was $15.03 while the total price for the non-local counterpart was $16.91" (i). The differences observed were not statistically significant, but they failed to reflect consumers' common assumptions about local foods. Another study, conducted for the Northeast Organic Farming Association, went beyond a comparison of local and non-local foods (Claro 2011). Rather, this study compared the prices of local organic foods with conventional grocery store produce. In this case, for the majority of the foods compared, the farmers selling organic products did charge a premium. However, it was not the overwhelming majority one might expect. Out of fourteen vegetables and fruits, four were actually cheaper when the organic variety was purchased at the farmers' market as opposed to the conventional variety at the supermarket. One of the items was the same price in both venues. In a comparison of organic food from the farmers' market and organic food from the supermarket, the farmers' market prices were lower on all but one of the fourteen items. Such findings hardly reflect the taken-for-granted assumption that informs so much of the popular conversation surrounding organic foods. Among small farmers who market directly to consumers, few are making a quick profit on organic. John and Katie certainly weren't. Nevertheless, they felt comfortable with their prices.

When consumers and commentators discuss how expensive organic food can be, they often are discussing retail price premiums—what one would pay at the grocery store for an organic banana or a pound of organic grapes. However, little attention is paid to farm-gate premiums—the premiums farmers are actually paid for their organic crops. Moreover, these discussions largely ignore the small organic farmers, like John and Katie, who have decided not to sell their produce wholesale to larger retailers. In other words, the discussion of organic food prices does little to explain the pricing decisions of farmers who seek to provide an alternative to supermarket organic.

What we do know about farm-gate premiums is that they are substantially more modest than those paid at supermarkets and grocery stores. The profitability of organic for the farmers is typically similar to that of conventional farmers; "There are exceptions to this rule, but the general pattern is one in which larger organic farmers achieve levels of economic performance that are more-or-less commensurate with their conventional peers, not substantially higher or lower" (Lockie et al. 2006, 86). Such data would suggest that high retail prices do not necessarily trickle down, leading to high-income farmers.

Pointing to the case of California salad green production, Guthman has identified one of the causes of this disparity, arguing that "price competition has ensued around other commodities in which agribusiness has expanded, as well, so that farm-gate premiums have virtually disappeared" (2004c). As a result, on organic farms that sell to conventional retail outlets, one would be unlikely to observe premiums commensurate to those paid by consumers at the grocery store. Forced to compete, farmers must sell for less. As a result, just like in the larger agricultural economy, farmers see little of the organic consumer's food dollar.

Such competition, and the siphoning off of profits by processors and distributors, may explain some of the disparity in organic agriculture between the cost of organic foods and the premiums farmers actually receive. However, John and Katie choose not to sell to conventional retailers, so no one is skimming off their profits—and they are not competing with agribusinesses for CSA sales. Their organic practices are not solely the consequence of structural changes within the organic sector. Instead, these practices should be seen as attempts to viably integrate structural constraints with the local conditions of the farm and with the vision of organic agriculture held by its farmers, farmworkers, and consumer networks. As we have seen, farmers work hard to balance their economic needs with practices that foster the types of community they hope to create. Throwing an extra item into CSA bags, padding bunches of kale, and treating volunteers to perks show the ways these marketing decisions are both reflections of and preconditions for particular types of social relations with the potential, at least, for being far more meaningful than those fostered by the industrial organic food system.

The sacrifices involved in ensuring a good match between the economic realities of business and the socio-ecological relationships of small-scale sustainable farmers are not limited to Scenic View Farm. Rather, we see such sacrifices in CSA programs across the country. Farmers like Katie and John want to treat their CSA shareholders as friends and members when they are volunteering—giving them an extra head of broccoli or a handful of patty pan squash as a neighborly thank you. However, when things go bad—when the tomatoes are blighted, the chervil bolts to seed during an unseasonably warm week, or the carrots get overtaken by weeds—they have a hard time not treating their shareholders as customers who deserve their money's worth. As a result, despite the claims that CSA programs help spread the risk of small-scale organic farmers among their members, DeLind suggests that few in fact allow members to bear the brunt of agriculture's inherent insecurities. Reflecting on her own experience of running a CSA, DeLind remarks, "We voluntarily exploited ourselves to give those who purchased shares a full return on their investment" (1999, 6). Even with marketing structures like CSAs designed to ease the burdens of struggling small farmers, countless farms like Scenic View continue to sacrifice in pursuit of a balance between genuine organic integrity and tenuous economic solvency.

If we are to consider small-scale local farms from the position of a reflexive politics in place and a relational theory of exchange—considering them to be bounded in a mutually constitutive way to structural economic logics and local relationships, norms, and values—we then recognize that the alternative qualities and transformative potential of such farms and their relationships are hardly assured. Rather, these qualities are produced by farmers like Katie and John, who sacrifice the economic gains associated with less alternative practices out of personal commitments to an alternative way of life and an alternative way of farming. These are farmers who worry about crushing debt, who want to be responsible to their farms and their customers, who show bug holes in their kale to customers as a badge of honor (or at least as a necessary evil), and who mean it when they say they want to give their customers a good deal. These are people who struggle to remain viable while farming in a way that leaves them even more vulnerable to the whims of wind and weather and the vagaries of pestilence and blight, as opposed to doing everything in their power to eliminate the capriciousness of "nature" (Goodman, Sorj, and Wilkensen 1987). These are individuals and families who have chosen an alternative way of life and are willing to struggle—to sacrifice in order to maintain good matches between their economic needs and ethical aspirations—to maintain such alternative agrarian lifestyles and livelihoods.

If we have reason to hope, it is because the small farms and committed consumers that make up our study and others like them demonstrate the continued viability of alternative visions of agriculture, albeit often fragile and imperfect, through the good matches they achieve. But at what cost? Tremendous sacrifice is often required of small-scale organic farmers in order to arrive at a good match that allows some level of economic success—or at least survival—while maintaining the beliefs and practices of the organic movement. Given such costs, what can be done to improve the future of the sustainable agriculture movement—to make it more transformative, sustainable, equitable, and just? In the following chapter, we direct our attention to the complex negotiations that allow small farmers, like those at Scenic View, to create organic futures—finding a farm-appropriate equilibrium between environmental ethics and the reality of their being a business.

seven NO-NONSENSE ORGANIC: NEGOTIATING EVERYDAY CONCERNS ABOUT THE ENVIRONMENT, HEALTH, AND THE AESTHETICS OF FARMING

On a late summer afternoon at Scenic View, while most of the workers were picking tomatoes, a speckled fawn wandered into a nearby field of summer squash. While several of the workers delighted in how cute it was, John took off chasing the animal like a hound, baying wildly. Even Katie seemed a bit alarmed by his response. Later that afternoon, when Katie sat down to weed a bed of onions, it seemed as though she had something on her mind. Finally, she came out and said it. "You would think, with how into the whole organic thing John is, he would be a total environmentalist," Katie sighed, still weeding between the rows of onions. "But he's not really. . . . I am always getting on him about filling the landfill with plastics. He says that's what it's there for. And he doesn't think twice about driving his car everywhere." Katie was being honest; some things on the farm weren't that green. The equipment all burned diesel fuel, and they traveled from field to field in a gas-guzzling pickup truck. Katie regretted the reliance on fossil fuels and wished they had an electric golf cart to make trips across the farm. For now, however, the farm was reliant on petroleum.

In such reflections we can see the ways John and Katie have negotiated matches among their lifestyle concerns, their environmental ethics, and their business as farmers. They draw a hard line around the organic label requirements that they will not cross—they can farm without pesticides and still produce food they can be proud of. However, they engage in other practices that are less than "green," like driving diesel-fuel-burning farm equipment, but that they see as necessary to their operations—at least for the time being.

As organic farmers, John and Katie were dependent on natural processes to make a living. Ladybugs served as pest control, bacteria were used to prevent fungi, and care was taken not to harm the natu-

Figure 7. Practical Environmentalism: Through sustainable management practices, farms like Scenic View are home to a variety of flora and fauna, like this autumn meadowhawk. (Photo courtesy of Connor Fitzmaurice.)

ral systems on which the farm depended. While perhaps not text-book environmentalism, it was, at the very least, an instrumentalist environmental ethic that was present on the farm and common among agrarian ethicists. Recall when every potato flower had to be picked by hand to prevent the bees from ingesting the bacteria used to kill the potato beetle larvae. Such actions undeniably illustrate an awareness of the farm's critical connection with the broader ecosystem. Whether the care was taken to protect the bees out of a respect for their intrinsic value or merely out of an awareness of their utility for the farm may be irrelevant; in the end, the bees were protected.

Maybe Katie was right. For John, nature was often a nuisance. Bugs were crushed between fingers, and families of woodchucks had their burrows flooded with water hoses. John may not have been an environmentalist, at least not in the stereotypical sense. He was a farmer. As scholars have noted, "The instrumentalism inherent to agrarianism"—even the most ecologically concerned visions of agrarianism—serves to prevent its consideration as environmentalism (Major 2011, 51). Environmentalism and work are often seen as diametrically opposed in the ways these concepts have been constructed: "For many environmentalists, work itself . . . requires and sustains the destruction of nature" (Moore, Pandian, and Kosek 2003, 7). Yet it was impossible to deny John and Katie's genuine connection and concern for their land via their physical engagement with it.

John and Katie aspired to an ecological agrarianism where environmental concerns were expressed through the practical linkages among people, land, and lifestyle (Major 2011). For the farmers in our study, both in the fields at Scenic View and across New England, forging practical, viable linkages between their businesses and their commitments to the environment, to human health, and to producing healthful agricultural products they could be proud of proved to be powerful forces shaping the everyday experiences of farming. However, much like Katie's evaluation of John's environmentalism, these were not expressions of pure ideals developed in a vacuum. Instead, concerns about nature, health, and the importance of producing quality products were ideals worked out in practice within the confines of life's challenges and opportunities. Just as market-related decisions, like pricing and sales strategies, require careful social negotiations, so do ethical concerns.

Working from the Ground Up: The Lived Experience of Organic Farming

One day, while eating lunch in the shade under a hedgerow of small trees on the margins of the tomato fields, John explained that he did not currently include a fallow year in his crop rotations since the land was needed to keep up with demand. He knew that many conscientious organic farmers included a fallow year among their practices—a principle going back to the founders of organic farming discussed above. However, John admitted that even if he added additional fields, he would have trouble maintaining a fallow one since it wouldn't be directly making any money for the farm.

"It's hard not to pay attention to money," John conceded, "because that's how we have managed to stay in business for all of these years." With a chuckle, he looked over to Katie and added, "That is, to the extent to which we are actually *in business*. I think we may finally be getting to the point where we're in business, don't you think?"

"We sure are *doing* a lot of business," Katie offered.

The joking banter revealed a seemingly constant tension behind the organic practices on Scenic View Farm. Despite John's joke, a six-acre farm with a one-hundred-member CSA and $80,000 a year in sales is certainly a business. However, the laughing covered up John and Katie's discomfort with the idea of farming being their business and not their lifestyle. In the day-to-day life of their farm, these tensions were mediated by the principles shaping their business practices. Sometimes, the practices weren't as sustainable as they could be; sometimes the pressure to stay afloat or to live a comfortable lifestyle meant that corners were cut. But the intentions were honorable. It's true that these were people who felt that leaving a field fallow just wasn't feasible for their farm because empty fields don't make profits. But these were also the very same people who refused to make use of the National List's exemption for copper sulfate in a year stricken with a historic blight. In other words, these were real people, neither textbook cases of ideology nor mere distillations of economic greed. But even financial derivatives traders have trouble living up to the cold calculability of *Homo economicus*, their economic rationality both constructed and bounded by their relationships and social commitments (MacKenzie and Millo 2003). The pragmatic environmentalism found on a farm like Scenic View consists of a careful balance among farmer values, ethics, and concerns about the bottom line.

Such negotiations among economic, environmental, and lifestyle concerns hardly fit with the linear model of a conventionalizing organic sector. Instead, they point to the "contested worlds" of contemporary organic farming (Rosin and Campbell 2009, 40). There are different sorts of organic markets with different practices and relationships—different sorts of human connections that are negotiated in fields of economic exchanges, subject to divergent geographies. Local, organic food sold via CSA shares provides a different market from the organic food sold by corporate food chains and a different vision of what organic could mean for future generations. Farmers also make a host of other negotiated matches—all of which give rise to different visions and meanings of organic practice. Understanding the organic sector—and its potential to provide a sustainable, organic future—requires understanding, among other things, how farmers match market and movement concerns in determining what practices they deem to be sustainable, sustainable enough, and off limits. In some ways, a farm's good match marks the line each farm draws in the sand for itself—the line that says, "This far, but no further." Regardless of the match attained, sustainable farmers are trying to achieve a relationship between economics and environmental ethics that fits their life choices and their local farming circumstances.

Good matches are being made on organic farms across the United States as organic farmers negotiate balancing the financial realities of an often precarious line of work with the deeply personal aspects of their organic commitments. Joel Rissman, who operates a diversified vegetable, grain, and livestock farm in Illinois, describes converting his uncle's farm to organic after realizing that conventional corn production wasn't working for him, his family, or the sustainability of his farm. Joel describes telling his uncle about his decision to transition to organic farming: "I told him it wasn't going to be easy, and it hasn't. But it is a lot more fun. You don't have to worry about rinsing [after pesticide applications] and your kids getting into something" (cited in Duram 2005, 117). While Joel views his farm as more profitable than his neighbors' conventional farming operations, Joel shouldered heavy burdens to do what he thought was right for his family and his farm. It wasn't easy. He isn't farming organically merely to capture the economic rents inherent in the organic certification system. Rather, his diverse motivations have led him to farm in a way that is both economically and ecologically more sustainable—and even more fun.

We sell such farmers short when we assume the entrance of agribusiness into the organic market undermines the ability of all small farmers to continue to engage in agro-ecological farming practices. The precise nature of the match certainly varies. For example, for some of our New England farmers the match was official organic certification. For others, like Collin at Viridian Farm, it meant practicing agriculture in accordance with the USDA regulations while choosing not to be certified since the farm could not justify the added cost when all of its business was direct to consumers. For others, an Integrated Pest Management regimen was viewed as providing the flexibility needed by their farms to prevent the total loss of a crop in the event of a disease outbreak—like the late tomato blight that struck New England in the early summer of 2009. Even for farmers like Sean at Verdant Acres Orchards—who viewed organic as a "farmers' market saying"—there was still an effort to find a match between economics and personal values. For him, being a good farmer entailed providing consumers with beautiful food that tasted better than anything they could get in a grocery store. In his own way—though quite differently from the more agro-ecological farmers with whom we spoke—Sean wanted to do right by his customers.

Any analysis of how these individuals understand ecological sustainability, the connections between food and health, and what a job well done looks like at harvest must begin with their everyday experiences—what Berger and Luckmann call "reality par excellence" (1966, 52). Here, we may understand how they construct the reality of "organic" through their interactions and social realities. In a way, it often appeared as though the individuals at Scenic View Farm were "habitually" organic. Asking John why he was a farmer—or, more important, why he was an organic farmer—was a lot like asking why a seed sprouts. There is an explanation for each of these circumstances; however, we are so accustomed to their realities that offering an explanation seems foolish. Seeds just sprout, and John just sees himself as an organic farmer. Over time, the ethics and economics of organic farming have been internalized as "schemes of thought and expression," or what sociologist Pierre Bourdieu (1977, 79) has called *habitus*. These schemes of thought serve as a "structuring structure," shaping the very types of farming practices John and the others conceived of as possible in the pursuit of a sustainable lifestyle. On a farm like Scenic View, where the daily reality of farming practice was far more pressing than abstract notions of sustainability, a bug hole on an otherwise perfect leaf of arugula stood as both a sign of a job well done and a reflection of ideological commitments so often left unarticulated in the workaday life of farming.

Even the character of Scenic View's farming practices as "certified organic" was largely taken for granted. And so it is with many aspects of farm life. As Michael Bell explains concerning the Iowa farmers he studied, the nature of farming is such that "these people are really wrapped up in what they're doing" (2004, 122). As a result, many of the big issues of agriculture are rarely consciously addressed. Bell suggests, "It may often—or even usually—not matter that the structure of agriculture in Iowa undermines economic, social, and environmental sustainability. . . . Sometimes it does matter to them, and very much so. . . . But the struggle to create a stable identity against the backdrop of uncertainty becomes an ongoing accomplishment that provides meaning to the lives of Iowa farmers" (ibid.).

These larger issues matter to the farmers at Scenic View, as well as to people on the other farms across New England in our study. However, concerns about larger issues—ethical commitments to sustainability, for example—were rarely directly confronted. And they hardly needed to be. For the farmers at Scenic View at least, they were integrated into the taken-for-granted practices involved in the daily life of the farm: an organic habitus. Wrapped up in the daily struggles of keeping an organic farm afloat, they mattered in a practical way.

Many of the everyday occurrences shaping farming practices at Scenic View, while of tremendous consequence for John, Katie, and their workers' experience of organic agriculture, are seemingly unremarkable. Most fundamental was the routine cycle of work during the week. Mondays were for morning harvesting of hardy vegetables and afternoon cultivation, Tuesdays were for picking the rest of the vegetables needed for Wednesday's CSA pickup, Wednesday mornings were for packing brown bags full of vegetables and sorting them for CSA pickup and delivery, and so on for every day of every week during harvest time on the farm. Of course, unexpected restaurant orders or a sudden encroachment of weeds could alter these patterns. However, unless such a deviation occurred, there was little question of what tomorrow would bring. It was never a question of whether to cultivate the vegetable beds; it was just a matter of what vegetables would need cultivating.

Equally obvious were the routines critical to the farm's commitment to organic practices. For example, when weeds overtook the rows of green beans, someone would grab a hoe. When hoeing did not work, the farmhands would get down on their hands and knees and begin to pull out the weeds. There are, of course, other ways to manage weeds. However, for the individuals at Scenic View Farm this particular set of practices had become taken for granted—obscuring the ways such seemingly mundane actions represent negotiated matches, balancing the economic, ethical, and sentimental components of the farm's agrarian relationships. Repeated season after season, these sorts of habits came to serve as the farm's objective norms of behavior. "The 'there we go again'" became the "'this is how these things are done'" (Berger and Luckmann 1966, 59). The discomfort experienced by Katie after a bad allergic reaction to parsnip greens, for example, led the farm to systematically avoid planting them year after year—despite growing consumer demand.

While no one would fault Scenic View for its lack of parsnips, some of the experiences and commitments shaping what practices became customary had profound consequences for how the farm approached sustainability. Animals were conspicuously missing from Scenic View Farm, despite the fact that they are central to creating the types of closed systems valued in discussions of agro-ecological sustainability. One afternoon, the subject of keeping animals came up. It had been a perfect summer day on Scenic View Farm as Katie, John, and their farmhands stood, stooped, and squatted, picking the last of the sugar snap peas off of their trellises. The conversation that afternoon had been lighthearted, if not jovial. Just before, John had been joking about the slow pace of everyone's picking.

"If I was a *real* farmer," John said, trying to keep from chuckling as his mouth turned up into a wry smile, "I'd make you pick with both hands." As it was, everyone kept one hand free for snacking. Clearly, no one needed to sneak a taste. But the peas, warmed by the summer sun, felt like the guiltiest of pleasures. The work was tedious but at the same time had been enjoyable. Suddenly, however, the mood soured. Katie was talking about wanting chickens for the farm. It was a seemingly innocuous desire but one that brought up painful memories, making John adamantly opposed to the idea.

Still picking, one moment kneeling and the next standing, Katie went on to describe the reason for his reticence. Years before, the farm had raised heritage pork for an upscale restaurant. One of their sows was pregnant, and everyone on the farm was eagerly awaiting the new arrivals. However, there were complications, and without time to bring in a veterinarian the sow had died in labor.

"I was devastated," Katie recalled, knowing that under their care, a mother hog and her offspring had perished. They never kept livestock again. The trauma of that event had become a structuring structure at Scenic View. Especially now, with their two-year-old son on the farm, John didn't want any more tragedies.

The practices of Scenic View Farm are forged out of efforts at brokering differences among an ecological agrarianism, the financial realities and pressures of the local food market in which the farm participates, and even deeply personal experiences of farm life, such as the loss of the sow so many years before. However, a myriad of other factors also shapes the ways the farm negotiates the practices that will be a good fit for it, supporting the farmers' vision of the type of farm they hope it will be. Besides environmental ethics, beliefs about health also informed the farm's commitment to organic practices. When asked why Scenic View was organic, John once responded that he just never liked the idea of spraying poisons on everything and then having to live among them. Katie elaborated that not only did she have little interest in eating pesticides, she also had no interest in feeding them to anyone else. As a result, beliefs about health and safety proved to be important motivations for John and Katie to farm organically.

Health concerns motivated several of our other New England farmers, who also avoided the use of chemical inputs on their farms or at the very least used them sparingly. Claire, for example, described her interest in working on an organic farm as stemming from her concerns about people's health and wellness. "I'm just really into health," she explained. "I'm actually going to school to be a certified health coach. So it's like working from the ground up. Working on a farm, I get to see the whole process of where our food comes from and having that connection. And I think the whole thing with our society is we're so disconnected from our food." Similarly, Mary at True Friends Farm studied wellness and alternative medicine while attending college, and she believed that "we really are what we eat" and that farming "fits really well with what I like." Elaine of Pioneer Farm lacked a formal background in health, yet she felt that it was important to make organic produce more affordable because more people should have access to safe, healthful foods. For such farmers, working in alternative agriculture fit well with their beliefs about the health and safety of our food system.

For some, concerns about health and safety were related directly to the visceral experience of farming. Donna, one of the workers at Scenic View, had never been involved with farming. As a result, much of Donna's outlook on sustainable agriculture developed through her participation in the practices at Scenic View. When Donna was asked about her beliefs about organic farming, the impact of the practices on the farm in shaping her perspectives was clear. For example, throughout the day, everyone in the fields would nibble on the crops and edible weeds with which they were working. Sure, one would get some of the gritty soil in one's mouth, but no one had to worry about pesticides. Commenting on her commitment to organic farming, Donna cited this practice as an example of why she valued organic agriculture. As a counterexample, she mentioned how a friend who had worked for a day on a conventional berry farm drew stares of disbelief after putting pesticide-laden fruit into his mouth straight from the field. Donna loved the freedom of not having to wash anything for fear of eating toxins; washing was necessary only if one didn't like the feel of sand between one's teeth.

Apart from health and safety concerns, a commitment to quality also played a role. As farmers, John and Katie wanted to produce the best food possible. Every farmer and farmworker we spoke to expressed a paramount concern to produce the highest-quality food. Some farmers understood quality very differently from John and Katie, who were willing to have a few bug holes in their leafy greens if it meant not spraying them with toxins. However, even Sean at Verdant Acres Orchards, who felt that it was impossible to grow good-looking food like that produced on his farm without using sprays, noted that providing customers with high-quality food was a major concern for him. "When you're bringing fresh tomatoes, fresh peaches, from out in the country and into the city, people are going to buy [the produce] because it actually tastes like something. We pick all our produce fresh and ripe. If you go to the grocery store, they pick it green. It isn't any good like ours; it has no flavor." Each of the farms in our study interpreted differently what it meant to produce the best food possible. For some, it meant food grown with organic methods. For others, low-spray methods allowed them to strike the right balance. Even farms that did not practice organic methods, like Verdant Acres, were doing what they thought was best to give customers a quality product. "If we didn't spray," Sean asserted as he gestured toward his farm's bountiful market displays, "we wouldn't be able to grow all this beautiful food." For others, a beautiful bunch of kale had a few labyrinthine markings from leaf miners, and a good ear of corn might need to have worms cut out. Surely the farmers might have hoped for a few less bug holes from time to time, and John and Katie often left countless turnips and radishes lying in the fields when their skin was riddled with the tunnels of grubs. Often, however, minor cosmetic damage was proudly displayed to customers as the objectified result of their organic practices—an aesthetic of sustainability.

One day, admiring the mountain of greens being given a triple wash behind the barn at Scenic View, Katie remarked, "If we can grow food this good and this beautiful without chemicals, what reason is there not to?" The 2009 growing season should have provided at least one possible reason, given the outbreak of tomato blight severe enough that it made the national news. And then there were the difficulties the folks at Scenic View faced year after year: potato plants overcome by potato beetles, cucumbers grown under mesh canopies, entire rows of chervil disappearing after having the life sucked from them by aphids. However, to those involved in organic farming at Scenic View, such pitfalls did not matter. Although problems such as infestations of pests or threats of disease often required creative solutions, they could not shake the taken-for-granted nature of the farm's organic identity. So after battling tomato hornworms each year by simply removing them by hand and crushing them under foot, John and Katie decided to plant buckwheat between the rows of nearby squash. The hope was that the buckwheat's flowers would attract parasitic wasps to lay their eggs in the hornworm caterpillars.

Such decisions built upon the organic values of farming with nature, rather than against it, as expressed by Ikerd (2001). And yet ideas about nature are also a means of relating to broader discourses and structures, not just local-level conditions or personal and idiosyncratic longings. Instead, constructions of nature and its cultivation reproduce social orderings of worth. As Moore, Pandian, and Kosek argue, "Neither ancestor of nor victim to human toil, nature itself is a means of enacting, expressing, and reproducing the works of humanity. Natures are made and manifest through embodied activity;

notions of nature work as discourse and ideology; and natural bodies are sustained through repeated material and symbolic practices" (2003, 8). Through cultivation, as these authors say, individuals shape not only the landscape, but also these same social orderings and their own identities. Organic farming techniques were not used because they worked the best in a production-efficiency sense or because they were the easiest methods at hand. Rather, as John described in the barn that rainy afternoon, Scenic View was "not going to be conventional, no matter what," because he and Katie saw organic practices as the right thing to do. For them, it was also a right way of being in the world. When crops are destroyed, the norms governing practices are far from practical in a purely economic sense. However, they remained the moral, habitual choice for those seeking an organic agrarian lifestyle. This is a personal choice, a valorized quality of life chosen over unfettered economic gain (Duram 1998).

Scenic View Farm's commitments to the environment, health, and a particular set of aesthetic concerns became rooted in seemingly objective, taken-for-granted practices—practices that shaped the ways the two younger interns thought about the farm as well. Since they were relatively recent additions to the farm, they were largely unaware of how the farm had become what it was. They did not fully understand why John had become a farmer; it was simply his role in life. For them, the organic character of the farm was a given—an objective social reality. Maggie and Joey never asked John and Katie why the farm was organic. As a result, when we asked, they were just as interested in knowing why. They had their own ideas, but they were not certain. Maggie, for instance, suspected that John and Katie were organic farmers because they were "just good people." She went on to cite all of the help they had offered her in the past and their daily demeanor. However, the fact remained that she was still curious. For nearly all of the farmworkers with whom we spoke, the tendency to take their farm's commitments to sustainable practices as a given was repeated.

While these individuals could describe why *they* thought sustainable and organic practices were important, few had ever been told—

or asked, for that matter—why the farmers they worked for decided to use such practices. Instead, concerns about the environment, human health, and quality were translated into undisputed, everyday practices of farm work through the relational networks and interpersonal commitments in which such values arose and were enacted. The organic farming practices at Scenic View represented environmental and ethical decisions, but they were decisions created in a community of routine practice producing pragmatic approaches to the environment, health, and quality rather than unmistakable ideological commitments.

Networks for Organic Farming

As we step back to look at why Scenic View Farm was organic and why these individuals were farming at all, it is probably important to remember that John did spend many of his early years in a farm context, visiting his family's farm during his childhood summers. Looking back on those years and telling his family's history, John described his grandmother as an early "organic farmer" who avoided pesticides on her fields long before it was fashionable. These early experiences undoubtedly shaped John's desire to be a farmer, structuring his patterns of thought and expression. John believed that it was these early experiences of visiting the farm that instilled the love of the outdoors in him and led him to the farm as an adult. However, John's commitment to organic farming also undoubtedly has much to do with his later experiences, particularly those with his farming mentor.

John's mentor was often mentioned on the farm as an influence on its farming practices and philosophy. When John first determined that he would work his family's land and attempt to make a living— and a lifestyle—out of organic farming, he quickly became connected with a more experienced farmer who helped him connect with a regional organic farming association and certifier. John's views about organic certification, discussed in the previous chapter, could become taken for granted in the context of his interactions with older, more established members of the alternative agriculture community who likely remembered the long fight to gain legitimacy and recognition from the USDA. John's mentor has remained an important part of both John's business and personal life, as he continues to farm nearby. And they still help each other out. When the particularly labor-intensive work of hand digging potatoes had to be accomplished, in previous seasons John, Katie, and their workers would lend their hands on his mentor's farm. Such mentorship relationships are not uncommon, although they can often become fraught given the time commitment they involve. As Duram found with one Florida farm, such relationships can become competitive after years of intensive mentorship. Nevertheless, given the dearth of state investment in alternative agriculture practices, such relationships are also vital. As one of the farmers she studied related, "You basically have to go to other organic farmers" (2005, 176).

For any new farmer without a family history of agriculture, lack of experience can be a significant handicap. Moreover, new organic farmers experience additional hardships since agricultural extension services tend to be geared toward conventional farming methods. As a result, mentoring is critically important for beginning farmers in the position John was in when he first started out. As a result of "the paucity of information about practices that deviate from the dominant agricultural system, farmers and rural advocates from Arkansas to Maine to the state of Washington have organized a growing number of locally based farmer networks during recent years" (Hassanein 1999, 2).

In overcoming the general lack of information about alternative farming practices available from the USDA, farmers like Joel from Illinois have relied on such networks—and specifically mentors. Joel explained, "I go to all of the meetings. . . . There is this one man I call my organic coach. I go to the meetings early and find him, sit down, and tell him what I am doing" (cited in Duram 2005, 176–177). However, not all farmers view cooperation in the same way. Some farmers, particularly those who sell to wholesalers, are reluctant to share their keys to success. Mary, a citrus grower in Florida, explains:

"People take up years of our time. . . . We don't mind helping you. But listen, this is getting a little old. And as soon as they get certified, guess what they do? Stab you right in the back, and I don't have time for that" (cited in ibid., 177). Such competition, however, never seemed to be an issue at Scenic View Farm. John's mentor was his organic coach. Even after years of farming, John still spoke about him.

The importance of these types of networks on the practices of Scenic View Farm was confirmed after John explained why he marketed using a CSA model. Again, the critical role of networks in transmitting knowledge about sustainable practice became apparent. John described how he had been running a CSA since the very beginning, starting with merely ten members. What is important to note here, however, is how he began using the CSA model: "I had gone to a regional organic farming conference, and they had a workshop on how to set one up. It seemed like a good idea, so I went back and started one."

Such relational exchanges are equally important in accounting for how John and Katie's farmhands came to see their jobs as valuable work experience—if not the start of a career. Joey and Maggie had both gone to college at a school with a large, student-led sustainable agriculture initiative, and such an initiative helped shaped their interest in farming and their outlook on sustainability. Joey had even been involved in running an elective class dedicated to teaching practical organic farming methods while he attended school. Institutions act as brokers when they "both intentionally and unintentionally connect people to other people, organizations, and their resources" (Small 2009, vi). Such examples demonstrate the ways institutions involved in supporting alternative visions of agriculture can broker relationships and provide resources for their members. Such brokerage can obviously have powerful results, connecting students to resources—such as the acquisition of land for starting elective courses in sustainable agriculture—and helping small farmers connect to the resources they need to become certified as organic, start CSAs, and find mentors (Thompson and Press 2014).

With these sorts of institutions and relationships, it becomes much easier for ethical commitments to be transformed into durable habits of mind, practice, and economic exchange. One afternoon, while weaving the fruit-laden branches of countless tomato plants into their trellises, Joey explained why he was involved in organic agriculture at Scenic View Farm. After all, Joey had left college to pursue farming. Since this was undoubtedly a big decision, one might expect a well-articulated, well-considered, perhaps even deeply ideological, answer. Surprisingly, he merely offered that food was important to him because "you are what you eat." A cliché truism.

Yet it is more profound an answer than it first appears. The following week Joey was thinning the numerous rows of beets, a tedious task given the nature of the plants. Beet seeds are, in essence, small fruits containing up to seven individual seeds. When they sprout, the young plants are literally growing on top of one another. The task of the farmer is to gradually reduce the plants down to one per seed, allowing the beet's roots the necessary room to swell in size. Sitting in the dirt pulling the weakest of the young plants, Joey began to talk. "Remember when you asked last week about why I was here farming?" he asked. "Well, I realized I hadn't given you a very good answer because, as strange as it may be, I never really thought about it before. I've done some thinking and I think I have a better answer for you if you're still interested."

Joey went on to elaborate on the notion of "you are what you eat." He suggested that societies are what they eat, with many social problems like rural poverty and especially environmental degradation stemming from problems of agriculture. He further described how organic farming was a way of addressing such problems—since food stands at the intersection of so many social systems—and that he felt as though sustainable agriculture could be a quiet, productive revolution. As a member of the sustainable agriculture movement in college, Joey was able to see organic agriculture as more than a possible career opportunity; it was the right thing to do. Finding himself in social contexts that legitimated this belief, both at school and on the farm, the need to critically analyze why he was choosing to act the way he was became unnecessary. He could afford to take these choices for granted because everyone else around him did. In the end, the paramount reality of daily life on the farm took on such a taken-for-granted quality.

Habits become deeply ingrained in individuals, and the organic character of Scenic View continues to be created out of the everyday experiences of the individuals involved in the life of the farm. Just as Katie's painful experience of parsnip greens served to determine what crops would be grown each season, so too did each individual's experiences of farming and interacting with nature, with growing and eating food, shape the reality of Scenic View Farm as organic. The views of John's good friend Mark about the farm's organic practices demonstrate how the carefully negotiated matches of organic agriculture come to be taken for granted through the relationships and social commitments of farm life. When asked whether he would have stayed at the farm for so many years if it hadn't been organic, he responded adamantly, "No. But I knew from the beginning John would never [not be organic]."

John could have changed his mind and begun using pesticides on his farm to increase crop yields, and he might have been tempted to do so during the hard times of pest infestation. However, through their long-standing interactions, John, Mark, and the others on the farm had come to trust the farm's organic characteristics as given. Through the types of ongoing, meaningful relationships Scenic View fostered, Mark could feel that he "knew" that John and Katie would never be otherwise than organic. Through their interpersonal negotiations, these individuals could settle on shared meanings, values, and boundaries to distinguish their economic work from conventional agriculture, conventional organic, or even a conventional business.

Despite how critical these types of experiences may be in an appropriate account of why individuals such as these choose to farm organically, their practices are also profoundly shaped by equally important relationships with larger economic, regulatory, and institutional contexts. However, even the organic regulations to which Scenic View adhered could, at least to some extent, be taken for granted, through framings that rendered them contextual constraints for organic practices and not as the bases of practices themselves. As a result, relationships with such structures become a critical part of the often unremarked-upon patterns of daily decision making for farms like Scenic View.

The Backstories of Organic Lifestyles: Organic Markets, Institutions, and Histories

Stepping back from the sights and sounds of farm life, we should understand the undeniable force of economic and regulatory constraints on individual and collective action at Scenic View Farm and beyond. The system of organic certification profoundly shapes Scenic View Farm since it is illegal to sell produce in the United States and call it "organic" without certification. According to the USDA's National Organic Program, "Agricultural products that are sold, labeled, or represented as organic must be produced and processed in accordance with the NOP standards. Except for operations whose gross income from organic sales totals $5,000 or less [annually], farm and processing operations that grow and process organic agricultural products must be certified by USDA-accredited certifying agents" (Agricultural Marketing Service 2008, 1). Regardless of the interpretations John and Katie attached to the act of certification and the meanings they attached to their lifestyle and practices, if Scenic View wanted to sell produce that could legally be labeled as "organic," they had to become certified.

There are also numerous unavoidable structural realities that go along with organic certification, and these powerfully influence Scenic View's farming practices. While these regulations have resulted in large-scale shifts in the contemporary organic sector (see P. Allen and Kovach 2000; Guthman 2004a; Lockie et al. 2006), here we are concerned with the ways these larger structural changes are matched to local practices to produce the types of organic identities that farmers imagine for themselves and their farms. Once Scenic View was certified, there was no going back without its losing access to the organic label. As a result, every action on the farm was in some way impacted. For example, when John and Katie found currants growing on a neighbor's adjoining land and their neighbor allowed them to pick them to sell, they could not call them organic since they were not from their property—even though no sprays had been used on the fruit and they were growing along the farm's property line. Moreover, they had to be very careful in handling the currants to maintain adherence to the NOP regulations. John and Katie would not even put them in the same refrigerator as their own organic produce since the organic guidelines require that organic and non-organic products be kept separate to prevent the accidental transfer of prohibited substances like pesticides.

Likewise, many other activities on the farm make sense only in terms of the objective, external rules governing their practices. For example, an important part of farming at Scenic View involved tools not often associated with working the land—a pad of paper and a pen. In order to maintain the organic farming certification, every significant activity had to be recorded. Whenever seeds were planted, the number, variety, and date had to be noted. Whenever an organic pest control was applied, it had to be detailed in a logbook. The same was true for harvesting and sales. This constant recording of the day's activities is precisely the bureaucratic wrangling that farmers mention in their explanations for why they choose not to be certified. However, since Scenic View is organic, this rule has altered the farm's practices. Katie, for one, rationalized the burdensome paperwork as a helpful way to enact the goals of sustainability that shaped Scenic View. Such an approach made the logbook a viable match with the type of practices she envisioned for the farm—for example, it allowed her to see what varieties were most disease resistant in the local conditions, what rotations seemed most beneficial, and what intercropping strategies mitigated pests. Katie argued that she and John often went back to the logbook for planning future years' planting, harvesting, and marketing schedules. However, despite this subjective impression of the practice, even Katie noted that they probably would not be recording everything if they did not have to.

Along with the structuring effects of rules, resources are an undeniable structural constraint on any farm. As a result, in order to fully understand why Scenic View Farm is organic, the structuring effects of resources on the farm's practices must be considered. The most obvious and important example of the paramount significance of resources is the capitalist economic system whereby resources are distributed in our society. In order to continue farming, Scenic View must succeed as a *business*. The capitalist structure of agriculture is the reason why we must ask how organic farmers seek to match their relationships to certain kinds of land, to particular kinds of markets involving particular types of social interconnections. It is the reason why so many stories from our fieldwork reveal constant attempts, on the part of the farmers and workers, to negotiate and match the competing pressures of love and money. In such an economic system, farmers—like everyone else—need to carefully match the types of relationships they hope to cultivate with their land, their families, and their communities with forms of economic activity that are comfortable for them.

While innovative marketing approaches like CSAs have helped farms like Scenic View survive and practice principled forms of organic agriculture, we still have a long way to go. Farmers like John and Katie are still struggling. Although they are doing what they love, staying true to their organic farming principles, and managing to get by, producing healthful food in ecologically sustainable ways shouldn't have to be so difficult. Just as the conventionalization thesis alerts us to the fact that all organic farms operate within the context of a larger political economy, the pressures of lifestyle serve as a reminder that such farms also operate within a cultural economy of norms, values, symbols, and aspirations. Farmers like John and Katie continue to fall short of achieving economic stability. Such farmers are already working incredibly hard to face the challenges of economic, social, and ecological sustainability. If farmers like John and Katie are to be able to continue farming in truly alternative ways, a new array of support mechanisms will need to be found and implemented.

While small-scale organic farmers often seek an alternative vision of the "good life," we have seen how difficult that can be. As we have already described, of more than two million farms nationwide, less than half are showing profits (Munoz 2010). In terms of John and Katie's values and work aspirations, organic farming would seem to be the perfect job. However, given the dire statistics concerning farm profitability, farming doesn't stack up so well against their aspirations for a comfortable life. To remain profitable as an organic farm while not cutting corners and intensifying production is a difficult task. Indeed, economic pressures have led to some degree of intensification on Scenic View; however, such intensification is not reflective of conventionalization. Rather, the type of intensification observed on Scenic View is by no means limited to organic agriculture. On the contrary, it is symptomatic of the contemporary American economy, characterized by Juliet Schor (1991) as driven by a work-spend cycle.

According to Schor, "Attitudes toward consumption come to be determined by the interdependent process of earning and spending" (1996, 48). For organic farmers like John and Katie, the pursuit of a comfortable lifestyle required abandoning some of the hallmarks of organic practice, like forgoing fallow periods for their fields and expanding the amount of land under cultivation. It is a cycle of work–intensify–spend, gradually eroding the outposts of alternative agriculture they have worked so hard to create. Nevertheless, Scenic View Farm has managed to remain remarkably committed to the ideals of sustainability. Working to achieve a good match between their values and the market, John and Katie have achieved balance between economic exigencies of staying in business and their ecological concerns and lifestyle choices. However, it is not without sacrifice. Lifestyle patterns and levels of consumption reflect class-based cultural norms and values (Bourdieu 1984; Schor 1998). As a result, remaining an outpost of alternative agriculture has meant that John and Katie have also had to strike a balance between conventional notions of "the good life" and the social and ecological values rooted in the historic organic movement they have adopted.

It is here that their stories also exemplify the challenges still facing alternative agriculture. Although the balance achieved on Scenic View generally seems to work—in that John and Katie have achieved a seemingly good match between economic rationality and their intimate connections with the land and community—they occasionally feel remorse that a comfortable existence appears just out of reach. When we first met John and Katie, we overheard them talking about grocery shopping while we were sitting down in their kitchen for lunch. When they returned to the table, John chuckled, "It's kind of lame that the people growing the organic food can't afford to shop at a place like Whole Foods." The fact that the work of providing healthful food while stewarding agricultural resources for future generations does not always pay off is a bitter pill to swallow.

The pressure that the broader economic context of American life exerts on John and Katie is all the more evident since farming isn't all that occurs at Scenic View Farm. Rather, the need for supplemental income to keep the farm operating has led to the farm's rental business. As we described above, the farm was fortunate enough to have a large house and a small cottage on the property that were made available for short-term—most often weekly—rentals. Since the properties were merely rented to guests without any additional services (as would be the case with a bed and breakfast or an inn), John and Katie were able to spend most of their time in the fields. However, despite the limited time entailed with these rentals, the supplemental income they provided was critically important.

The pressure to be profitable, or at least financially sustainable, was a constant worry. One afternoon while weeding between rows of red onions, Katie talked about the farm's finances. As we spoke, the conversation drifted to the rental business. "If it wasn't for the rentals," Katie said, "Mark probably wouldn't be able to make a living here. It's that important." Clearly, the reality of the farm is shaped in large part by the structuring of resources. Farming alone is not enough to sustain the ethical choices and lifestyle decisions that have brought these people together. Rather, the rental business—whether they enjoy having it or not—is necessary for Scenic View to continue operating in its current form.

Moreover, the rental business works only because of the type of agrarian lifestyle and relationships Scenic View cultivates. Such a rental business wouldn't be a good match for an industrial farm seeking supplemental non-farm income. For one thing, the natural landscape John and Katie inherited made such a business possible for Scenic View. The farm could have this extra source of business only because it is an undeniably beautiful place, nestled among forests and wetlands in a scenic region of New England. People *want* to rent space and get married on the farm. As a result, just as the need for resources has undoubtedly necessitated that the farm produce supplemental income, the ecological resources on and around the farm structure how that income can be generated. In additional, the rental business was critically linked to the type of landscape John and Katie actively cultivate through their farming practices, producing a particular agrarian aesthetic suitable for tourism. Such commitments to preserve a particular landscape reveal the ways farm practices reference and relate to forces beyond the personal connections of farm life— to structures of culture, class, and race. Landscapes, after all, are both a "way of seeing" and a product of the cultivation of what is seen (Moore, Pandian, and Kosek 2003, 11). The act of cultivation, and the discourses surrounding it, have historically been rooted in the production of racialized hierarchies of worth (ibid.). As a result, the moral worth attributed to organic landscapes cultivated by small-farm owners can serve to reproduce spaces of privileged vision and whiteness, eliding the oppressive nature of agriculture for nonwhite bodies throughout history (Alkon and McCullen 2011; Guthman 2008a). Once the rental business was established, it exerted a reciprocal effect on the farm's practices: intensifying agricultural production could spoil the value of Scenic View's landscape.

Beyond the landscape, the land itself was of paramount importance for making farming feasible for the individuals at Scenic View. It is important to remember that the property utilized by Scenic View Farm has been in John's family since the turn of the last century. This important resource was available to him when he decided to adopt his new career, shaped by histories of white land ownership, inheritance, and privileged access to loans—affordances systematically denied to nonwhites in the agricultural landscape (Guthman 2008a).

The importance of the availability of land cannot be ignored. The same afternoon Katie explained the importance of the rental business, she mentioned that they needed to rent the property from John's parents, albeit at a low cost, since his family still incurs costs maintaining the large property. However, the minimal nature of their rent payments made things much easier on the farm. Without this family asset, farming would have been much more difficult. Unlike most other farmers, John and Katie listed the cost of land as one of their least significant expenses, coming in after labor (their largest expense), supplies (including fencing and row covers), and inputs (items such as seeds, supplemental fertilizer, and lime). As a result, all of the reasons John gave for farming—such as his desire to work outside, his love for food, and even his desire to get back to take care of his family's land—must be considered in light of the fact that his family had the land in the first place. The availability of land formed the context for John's decision to pursue farming.

As we have already described, the cost of land is undoubtedly one of the largest barriers for new farmers in the United States, with the cost of farmland doubling over the course of the last decade. Emily Oakley, for example, who described the isolation of the rural community in which she and her husband ended up after their search for cheap land, had trouble even getting a loan for her farm. They had $25,000 saved when they approached a loan officer for a government farm loan. The officer laughed. Emily explained, "He'd never met anybody coming in for a loan for an organic vegetable production" (cited in Raftery 2011). For organic farmers who cannot afford to purchase farmland, renting can be equally precarious. It takes three years of organic management before a farm can be certified. If the owner decides to stop renting the land or the property is sold for development, those years of hard work are wasted (Weise 2009). In light of such uncertainty, the fact that John rents from his family and that the land cannot be sold for development since it is protected by a conservation trust take on tremendous importance.

Although the individuals' subjective interpretations of why they were farming organically at Scenic View are meaningful in and of themselves—and are certainly valid reasons to engage in such practices—in order to grasp the full social reality of their practices it is necessary to view them in light of the larger structural conditions that narrow the realm of possibility for all farmers across the region and likely across the United States. These conditions are not, in and of themselves, the reasons why these individuals farm the way they do, but they shape the many struggles the farmers face while simultaneously making it possible for them to succeed as organic farmers.

In the case of Scenic View Farm, it was clear that ethics played a critical role in shaping farming practices. However, enacting ethical relationships necessitated matching agrarian beliefs and values with appropriate forms of economic activity. These beliefs include an ethical commitment to the land and seeing organic farming as a way of nurturing the earth and soil so that it will support us for years to come. These beliefs also involve a commitment to living outside and not being constrained to an unfulfilling "desk job." Finally, these beliefs involve a commitment to the community with the cultivation of high-quality products free from chemical residues. Through good matches, farmers—like those at Scenic View—are using the market to enact these types of agrarian relationships.

On the other hand, by enacting these principles via matching them to appropriate forms of economic activity, they are necessarily tied to larger market forces beyond their direct control. Good matches require constant renegotiation to ensure that neither the sentimental nor the economic components of the relationship compromise organic integrity. "It's hard, though," John admitted. "You have to be concerned about money. It's not like we're trying to get rich. It's just that we have to stay in business so that we can keep doing what we love." Being in business remains both an inescapable reality and an effective means of living the type of organic lifestyle John has envisioned for himself and his family.

John and Katie's articulation of the complex balancing act of staying in business to keep doing what they love is echoed in Brian Donahue's account of his efforts to establish Land's Sake, a market garden in the suburbs of Boston. In the early years of Land's Sake, weeds overtook the main cash crop of strawberries, and Donahue and his partners almost lost their farm as a result of their refusal to compromise their commitment to avoiding chemical pesticides and herbicides. Remarkably, Land's Sake survived, and in Donahue's account we can see clear similarities to the types of negotiations central to the "good match" John and Katie have forged at Scenic View. Ecological concerns loomed large at Land's Sake, but Donahue also recognizes that for an alternative farm, "it would do no good to go broke, either" (1999, 80).

Donahue describes such negotiations of economic realities with ethical imperatives as "distort[ing] our own market calculus" (1999, 80), and such distortions are precisely why we must view the economic activity of organic farming as more then merely "embedded" in a socio-ecological system. The willingness of farmers like Donahue to nearly lose their farms as a result of their ethical commitments represents more than a mere external constraint on an otherwise rational market. Rather, the organic farmer's economic calculus is always fundamentally open to distortion by the complex realities of social, ecological, lifestyle, and ethical commitments.

As we consider why Scenic View Farm is a certified organic farm, consider that rainy afternoon in early July in the barn with Katie and John. The torrential rain pounding on the sloping metal roof gave one the feeling of standing in a tin can. Waiting for the last CSA pickups for the day, John and Katie related the stories behind the photos on the walls and played makeshift games of shuffleboard on the barn floor with a stone and a straw broom. That day led to many interactions with all sorts of people who valued the farm and its goals—volunteers, customers, and friends. These interactions make up the taken-for-granted daily reality of these individuals and are a vivid part of their daily lives. The practices of Scenic View Farm are motivated by commitments to ecology, health, community, and aesthetics. But they are also motivated by the pragmatic decisions of people and the way of life they imagine for themselves, the subjective agents whose ethical decisions constitute the reality of the place.

For John, Katie, and the others, those ethical decisions were often taken for granted: they were just common sense. However, to truly grasp why these individuals are farming organically, despite all of the struggles, the most taken-for-granted aspect of daily life—the structural context in which these organic farmers live—must be understood. Although the accounts of the individuals who made up the farm create the reality of the place, they do not occur in isolation. The individuals are responding to structural conditions (organic certification, regional conditions, diversity on and off the farm, access to land, etc.) that span the region and the country, affecting countless other farms like theirs.

While the structural conditions of a bifurcated organic sector might be similar for countless small organic farms, each farmer responds to those conditions based on his or her lived experiences, and each forges matches between those conditions and local practices in ways that mark the types of agricultural relationships he or she hopes to cultivate. These individuals' decisions to engage in organic farming are only a small part of their larger personal histories, which have equipped them with resources and dispositions allowing them to see organic farming as a possibility in their lives. Moreover, these people participate in the complex and often contentious social and regulatory history of the organic farming movement as a whole, shaping their conceptions of what organic agriculture is and can be.

Thinking about good matches helps to make sense of the nuanced ways in which small farmers respond to the seemingly impersonal structural forces that have led to the bifurcation of the organic market. The structural conditions in society both favoring and constraining organic farms, the regulations requiring certification, and John and Katie's access to resources such as affordable land all permit the various organic practices witnessed on Scenic View. But while these conditions may permit certain practices, individuals' interactions with each other, the land, and the larger organic market give meaning and shape to these practices. As a result, while the structural conditions permit John and Katie to sell only "organic" produce from a certified farm, Scenic View Farm is not certified organic simply because of this condition. Rather it is certified organic because it meshes well with the types of economic and agrarian relationships in which John and Katie hope to engage and the types of farmers they imagine themselves to be.

For some other farmers, achieving a good match between their ethics and their business model did not mean organic certification. For some it just meant a responsible use of pesticides and a strong sense of accountability to their customers. But for all of the farmers in our study, these rich, deeply personal, and even idiosyncratic explanations are critical aspects of why these farmers practice agriculture in the ways that they do and struggle to match such practices to their economic needs as best as they can year after year.

CONCLUSION: AN ALTERNATIVE AGRICULTURE FOR OUR TIME

> Farms still need to create healthy cultures in their soils, on their plates, and in their communities; this is a component of sustainability not included in organic certification.
>
> —Molly, Grateful Harvest Farm, 2013

Years have passed, and today Scenic View Farm is a different place. For one thing, John and Katie now have two children in tow as they weed the rows of baby carrot seedlings and pick the first tender leafy greens of spring. Matt has since left, having moved off the farm with his wife to support her in an important career opportunity in the food world. Maggie has departed the farm too, in order to advance her own career outside of agriculture. Joey has stuck around the farm, continuing to apprentice under John and Katie in a more managerial role in hopes of getting another step closer to a farm or nursery of his own one day.

It was the connections the farm and its workers had to the regional food world that led John and Katie to embrace sales to restaurants as a vital part of their business, while many of the other farmers in our study avoided such commitments, deciding that such sales were too demanding and limited for them to go out of their way to secure. It was these same connections that led to the job that ultimately took Matt and his wife away from the farm. Years before, it was John's relationship with his mentor that propelled his entry into farming. Today, John, Katie, and their workers still occasionally lend a hand on his mentor's farm when he is short on labor for difficult tasks. Now, years later, it is Joey's relationships at Scenic View that allow him to see a career in agriculture for himself. His experiences under John and Katie's mentorship have kept him working in their fields and have led him to accept greater responsibility for the farm's daily operations. Meanwhile, John and Katie's new daughter serves to reinforce their commitment to avoid harmful chemicals—even allowable ones—in their fields. They want their farm to be a place where they can raise a family and wander "the garden" burdened by one less worry about their family's health and well-being. Today, their son already knows the phrase "Let's go cultivate the chard," and their daughter is learning to walk—with as much delicacy as a toddler can muster—between the rows of bull's blood beets and speckled trout lettuce.

Like many ethnographies of farmers' lives and livelihoods, the previous chapters have been filled with examples of the importance of farmers' social, emotional, and ethical connections—as well as the resources and opportunities open to farmers and farmworkers because of their particular life histories—in shaping the agricultural practices undertaken in their fields. Ethnographic approaches, such as the one we have taken, allow us to observe and understand the lives of small-scale farmers. More important, they allow us to uncover the everyday experiences at the heart of how and why alternative farmers enter into and sustain alternative businesses, social relationships, and lifestyles.

As we step back from Scenic View Farm, the other small farms throughout this book, and the empirical findings of other ethnographies, we can begin to ask what these findings can contribute to the broader debates about the contemporary food system and the possibilities of achieving more sustainable alternatives in the future. Despite recurrent ethnographic findings about the importance of non-economic exchanges, they remain largely absent from the broader theoretical debates about sustainability, organic practices, and alternative agriculture within the bulk of agro-food studies. Bringing economic sociology into these debates, we can begin to theorize how these types of relational forces impact the practices of farmers and the terrain of agriculture's political economy.

Studies of economic relations experienced by small-scale farmers like those at Scenic View are somewhat missing from the relational economic sociology literature. As we outlined in the introductory chapter, economic sociology has largely set out to problematize the idea of separate spheres/hostile worlds in social and economic life (Zelizer 2010). As a result, the typical cases that have received attention are those in which markets emerge for goods and services commonly held to be outside the purview of rational economic exchange. The findings of the studies we highlighted in the introduction, for example, reveal how markets for services like sex (Hoang 2015) or goods such as organs do not inevitably erode the social significance of the products. In fact, the emergent markets are often dependent on the reproduction of those social meanings, as when organ donation proponents worked to reinforce beliefs about the sacredness of the body and the gift of life organ donations provide (Healy 2006). In small-scale alternative agriculture, however, there is a rather different set of social relations.

Unlike these other cases, the conventionalization of the organic sector suggests that in this particular market, the way economic rationality was introduced has limited the ability of some farmers to arrive at a workable match between organic farming practices and the meanings and relationships they attach to the idea of "organic" agriculture. In other words, some small farmers really do feel the flattening, homogenizing, and co-opting influences of the market. So they attempt to resist the dominance of the market, but not by falling back on the commonsense notion that ethical commitments and relationships of concern must be inevitably kept separate from market logics or suffer damage or dilution. Instead, some of these farmers are actively building new markets by cultivating the types of commitments and relationships that constitute so much of the character of Scenic View Farm. Without abandoning the idea of farming as a business, they work to foster more connected consumer experiences; they farm with practices that, while not always directly profitable, support their vision of organic; and they try to stay in business so that they can keep doing what they love. In other words, paradoxically, they try to build new, alternative markets to challenge the watering down of "organic" that the full-force entrance of market logics ushered in.

Moreover, bringing these types of social and ethical exchanges to the center of our analysis of farmers' lives has clear implications for how we think about the organic sector's political economy. In the daily workings and decision-making processes of Scenic View Farm, family histories, access to family land, childhood experiences in nature, connections to particular places, aesthetic ideals about what a farm should look or feel like, access to vacationers for side-business ventures, connections to the region's urban restaurant scene, and access to college students with backgrounds in environmental studies interested in putting their ideals into practice were just as important as the conditions of the organic sector's political economy—and were perhaps more acutely felt. Thinking about these connections, and the types of good matches individuals working in the food system can and cannot cultivate because of them, helps us better understand how alternative markets for organic products are created and exploited by small-scale farmers, as well as the types of relationships these farmers must either have or be able to foster in order to participate.

Many of the relationships Scenic View Farm needed in order to successfully take advantage of the niche market for small-scale local products must necessarily draw us back to concerns about the ability of alternative agriculture (even outside the supermarket produce aisles) to deliver a more holistically sustainable food system for our future. Scenic View was dependent on privileged relationships to land and bank loan officers, while at the same time struggling to foster the types of community involvement it desired and upon which it at times depended. Such matches (and failed matches) at least suggest that small-scale farms, in and of themselves, are not synonymous with true sustainability and a vibrant system of alternative agricultural practice.

While we have chosen to focus on the relational work small-scale farmers undertake to match their economic practices and business models to their relationships and lifestyles, thinking about the economic lives of organic farmers relationally means that improving sustainability inevitably requires more than just changes on farms and fields. It will also require new *social relations* of food *consumption*. The way agriculture was carried out on Scenic View was intimately tied to the ways the farm was able to relate to consumers—matching relations of exchange to its customers' expectations, demands, and levels of commitment to participating in the food system. Scenic View succeeded in garnering consumer commitment to purchase from the farm. Its restaurant clients were steady and loyal customers, knowing that Scenic View would provide high-quality, carefully harvested, hard-to-find seasonal ingredients. Its CSA subscription program was always full. It even attracted some volunteers, such as the folks who made packing one hundred CSA orders a mere morning's work. However, John and Katie still worried that maybe not everyone was as excited about connecting to food production as they were—making exchange relationships the dominant way they tried to connect to those who consumed the fruits of their labor.

For some, a system of agriculture predicated on small farmers "going it alone" to carve out outposts of alternative agriculture—for the privileged few to enjoy—is far too reflective of the problems of a neoliberalized food system than of a genuine solution to the economic, ecological, and social problems of our times. At the very least, the ways the politics of contemporary alternative food movements reinforce neoliberal market values would suggest that cautious consideration of the transformative potential of such movements is warranted. We live in a neoliberal era, where government support of public environmental resources, small-farm subsidies (not to be confused with the enormous subsidies that large farming operations receive through the Farm Bill), and other social programs are increasingly seen as wasteful expenditures. It is an era where the market is considered the place where all problems can be resolved, including environmental problems, and it means that opposition to mainstream markets (such as alternative food networks) is viewed as counterproductive or simply as yet another consumer choice. Rather, lean government, support for large-scale private enterprises (including large-scale industrial farms), and support for individuals operating in the marketplace are powerful political projects in a neoliberal world. As a result, the long-term survival of small-scale farms that rely on local communities is up against a lot of resistance in a world that supports large-scale, low-priced food. Critics of small-scale farms argue that these operations must either sell out their farming principles or go bankrupt. Or they must sell up—providing elite consumers with ethical, green, artisanal alternatives to mainstream agricultural products, leaving the rest of the food system behind.

Engaging with the Criticisms of Localism

The correspondence between many alternative agriculture initiatives and the dominant neoliberal approach to the food system stems in part from the very fact that the economic practices that many in small-scale agriculture see as good matches—such as organic practices and food localism initiatives—are not rejections of a competitive, market-based logic. The farmers we encountered at Scenic View tried to build the types of market relationships that could resolve the problems they saw in the larger food system and the economy in general. They sought economic arrangements that would allow them to farm in ways they could be proud of, have the type of lifestyle they desired, and foster relationships through their passion for food. More generally, formal labeling systems such as "organic," as well as more informal ones such as "local," serve to provide a competitive distinction among products (Guthman 2007) and support a system where economic benefits go to successful producers (McEntee 2011). In fostering consumer choice, however, some worry that these initiatives can only ever provide "partial and uneven" sustainability—for *their* farms, *their* consumers, and (maybe) *their* own workers (Guthman 2007, 473). As a result, many "projects in opposition to neoliberalizations of the food and agricultural sectors appear to have uncritically taken up ideas of localism, consumer choice, and value capture—ideas which seem standard to neoliberalism" (Guthman 2008b, 1174). While small farmers may be attempting to build businesses that match their agrarian values, some caution that the "solution" to a neoliberal system of industrial agriculture is increasingly cast in terms of small farms, unsupported by the state or their communities, producing food intensively on small parcels of land and competing for consumer market share—an equally neoliberal system of food production.

Moreover, many arguments for local food systems potentially conflate knowledge of a food's place of origin with a political consciousness "as if awareness of the intimacy of food will automatically propel one to make reflexive, ethical food decisions" (Guthman 2008b, 1175). Such unreflexive localism can lead to the unintended deepening of inequalities: an alternative food movement both produced by and reproducing the inequalities of a neoliberal food system (see DuPuis and Goodman 2005). Critics caution that "reducing the scale of human interactions does not necessarily achieve the social equity or empowerment espoused by alternative agrifood movements" (P. Allen 2004, 173). Instead, the creation of local, competitive markets could actually exacerbate local inequities for those left out of these alternatives by reinforcing the temporally and geographically unequal outcomes associated with the devolution of state control characterizing neoliberalism: "As a result of historical processes, communities vary widely in the resources they can bring to bear in developing sufficient, sustainable, and equitable agrifood systems" (P. Allen 2004, 177). As the case of Scenic View demonstrates, such resources can represent a critical tool for making the types of matches necessary for making a small-scale farm economically viable. As Patricia Allen aptly summarizes, "Globalization produces winners and losers, but so does localism" (179).

Outside the possibilities of localizing efforts exacerbating the local-level inequalities of neoliberalism, even "the increased salience of food politics in contemporary life may itself reflect the neoliberal turn, particularly insofar as much of what passes as politics these days is done through highly individualized purchasing decisions" (Guthman 2008b, 1175). While the assumption that local food systems can and do produce more equitable outcomes and empower food system participants is compatible with communitarian ideology, it is also entirely compatible with the libertarianism of neoliberal policies and politics (Harrison 2011). While the market-based approaches taken by the contemporary organic movement could represent efforts to build more equitable and sustainable economies, they could also be a "path of least resistance" since "some of the predominant forms of alternative agrifood activism have come to rely extensively on market mechanisms and [have] simultaneously abandoned the pursuit of regulatory reform" (Harrison 2011, 163).

Based on an understanding of social, political, and economic change in which consumers "vote with their dollars" in the marketplace, solutions to agricultural problems have come to be cast as dependent on "increasing consumer choice in the marketplace, framing problems as individual rather than structural, and dismissing the need for state interventions into industry activity" (Harrison 2011, 163). While (some) consumers may be able to choose more sustainable options, "obscured in the shadow of organics' spectacular rise remain the recurring pesticide drift incidents as well as troubling data about pesticide use of the other 97 percent of California's [nonorganic] agricultural land" (158). For Jill Harrison, then, the problem with solutions such as organic agriculture is that "the libertarian model of change is one in which social and environmental benefits are tied to isolated, individualized purchases, and thus are problematically undemocratic, contingent on ever-shifting consumer whims and abilities, enjoyed mostly by relatively privileged consumers, manifested unevenly across the landscape, and unbolstered by regulatory protections" (164). Such assessments would suggest that the approach taken by the alternative agriculture movement is tied to the same ideological underpinnings of the broader neoliberal paradigm in such a way that prevents it from countervailing against the agricultural problems of our times. In these accounts, alternative food movements, as they currently operate, are far from transformative. Rather, these critiques cast local, organic food as apolitical. In order to address the problems of contemporary agriculture, more critical approaches to local food systems would suggest that "food system localization will need to compel people to act as citizens beyond their purchasing decisions" (Harrison 2011, 167).

By putting the politics of our food system into the market, critics fear that sustainable foods will become nothing more than status symbols for those privileged enough to afford to support alternative agriculture. Organic salad greens became "yuppie chow" (Guthman 2003), as the consumption of local, organic, and artisanal fare became a distinguishing practice for the current generation of foodie trend-setters (Johnston and Baumann 2010). While the idea of voting with one's dollar might be attractive, the real world potential of such "in-dividualized" approaches to political action can be a mixed bag. Far from being a straightforward solution, consuming for change can obscure myriad ideological contradictions (Johnston 2008).

Chief among these contradictions—in terms of ecological, eco-nomic, and social sustainability—is the generation of "a cultural-ideology of consumerism, a political-economic denial of class inequality, and a political-ecological message of conservation through consumption" (Johnston 2008, 261). Certain organizations may be better able to "de-center the idea of consumer choice in the service of ideals like social justice, solidarity, and sustainability" (263) than can a supermarket like Whole Foods. More alternative retailers, such as food cooperatives, have sought to integrate concerns for social jus-tice into alternative food systems by supporting domestic versions of fair trade labeling, for example (Duram and Mead 2013; Upright 2012). Nevertheless, approaches reliant on consumers voting with their dollars miss "the failure of a neo-liberal mode of social repro-duction that idealizes markets and consumers as primary guardians of the public good, and occludes the importance of non-market mea-sures and democratically accountable states" (Johnston 2008, 263). Skeptics, who see small-scale organic farmers and the members of their communities who support them as building a model of con-sumer citizenship, point out that "while theories of transformative consumption are inspiring," there is a difference between "self-oriented consumerism and the collective responsibilities of citizen-ship" (Johnston and Szabo 2011, 305). These efforts could very well run the risk of being "part of, and potentially [supporting], a neolib-eral political culture that undermines a collective sense of civic re-sponsibility and state regulation of ecological issues" (317). If so, even a "reflexive localism" is potentially problematic; consumers often lack both the time and ability to "make the best shopping decision," and expecting them to do so amounts to "expecting reflexive consumers to regulate a complex food system" (317). But, we argue, there are other ways of considering the potential of alternative agriculture.

Digging Deeper: Social Consumption and Alternative Agricultural Production

Certainly scholars concerned with the affinities between neoliberalism and the solutions offered by the alternative food movement—such as localizing food systems, providing consumer alternatives, and capturing market value—describe genuine problems and challenges that constrain the possibility of a sustaining alternative agriculture in contemporary times. As the criticisms of efforts to localize food systems demonstrate, even niche market growers can be seen as reproducing neoliberalism's inequalities by focusing on consumer choice, existing in structural positions that complement rather than challenge the broader projects of neoliberalism in the food system. There are challenges and risks associated with alternative food movements advocating for decentralized approaches that build new producer and consumer relations.

While it is undeniable that emergent local food systems exist in a neoliberal context, their decentralized approach cannot automatically be assumed to be a part of the neoliberal project. Even some of those critical of the social and economic relations of alternative food movements recognize the practical and political appeal of localizing food systems: "On the practical side is an interest in reducing energy costs used in transporting and storing food," while the political "is based on interest in deepening democratic principles and practice" (P. Allen 2004, 165). Even if the form of many contemporary food movements potentially reflects and supports problematic neoliberal advances in modern farming, we need not write off all such efforts solely on the basis of apparent affinity. Rather, we must allow for the possibility that some of these efforts are alternative advances in an era marked by a government increasingly captured by business and unresponsive to its citizenry. Our investigations, in fact, show that the matches that local, small-scale growers forge in their agricultural networks offer some promising signs of radical counter-choices to the mainstream neoliberal model. Indeed, the efforts of small farms carving out outposts of alternative agriculture across the United States are not simply misguided or merely contributing to the problems of a neoliberalizing agricultural sector. Nor are they merely filling niches in which a conventionalizing organic sector isn't interested. Rather, we believe that the rich ethnographic data obtained through our case study, along with the vignettes of other small farms and farmers across the country, contribute to conversations about how to advance our understandings of the relationship between the alternative agriculture movement and neoliberalism. Such farms can potentially, *through the matches they forge with their practices and with their communities*, embody actors engaging in genuine attempts at countercultural agriculture.

In the fields—if not at the supermarket checkout—there are reasons to hope. The ways small-scale farmers understand and make sense of the often contradictory pressures they face as business people, community members, environmental stewards, and even as individuals have real-world repercussions and determine the matches they make with their practices. These matches shape their connections to their land, their communities, and their ability to embrace the type of agrarian lifestyle they value for themselves and their families. Such alternative visions, and the ways they match onto alternative practices, could form the basis for truly alternative market relations around food—not just a niche market. Given the realities of the contemporary agricultural sector—and contemporary politics more generally—we must allow for the possibility that such attempts at change could represent viable matches for the farmers making them and thus viable solutions for future organic agricultural networks.

Justification for making such allowances in order to recognize potential avenues of transformative change is, in fact, supported by scholarship on neoliberalism. One of the main problems with critiques of neoliberalism is that they can create a "there is no alternative" attitude. For example, it is nearly impossible to imagine a form of political resistance to neoliberalization that remains "uncontaminated" since "neoliberalism 'recognizes' political resistance as a performance of neoliberalism" (Bondi and Laurie 2005, 399). Avoiding such determinism requires recognizing that even neoliberalizing aspects of the alternative food movement could *potentially* be utilized to forge a more meaningfully alternative system. As a result, rather than farmers and food activists seeking out marginal areas in which to resist neoliberalism's advance, the core principles of neoliberalism may be its greatest vulnerability. The neoliberal notion that individuals can and should bring forth their own interests and engage in political contestation could be the very kernel that further popularizes small-scale food production networks that provide better, safer, more healthful food and a better sense of community than the isolating conventional shopping experience. And it is more realistic to view alternatives in this light because neoliberalism cannot be avoided by anyone. It must be confronted and dealt with and its flaws made more obvious to more people.

The local marketplace can be the site where such a confrontation happens, through the local networks forged through CSAs and other alternative markets, as can the rows of crops on small-scale farms. For many, it is the seemingly mundane "alternative practices of everyday life" that pose a great risk from within for the neoliberal era (Brenner and Theodore 2002, 346). While farmers like Katie and John at Scenic View Farm may not speak about the problems of neoliberalism, they are keenly aware of the constraints the current political and economic climate places on small farmers. Such constraints form the backdrop for their lives, and our ethnographic and interview data show that they are staking their lifestyles and livelihoods on doing something different despite the challenges. They are invested not only in alternative agricultural practices, but also in practically enacting alternative economic relations and lifestyles.

The many farms like Scenic View across the United States are undeniably enmeshed in a neoliberal agricultural context that other scholars have intricately mapped. However, in the fields we have witnessed that the neoliberalism of contemporary agriculture still contains alternative visions of economic life. In spite of trends toward conventionalization, some small farmers are able to resist, persisting in practices much more reflective of the more alternative principles of the organic movement than we should expect based on the most critical accounts. For all of the farmers to whom we spoke, the values of agrarianism suited them—and they negotiated between such values and their economic needs through their practices.

Organic practices were not always suitable for every farmer in our study. But the farmers with whom we spoke and the matches we observed all revealed a practical commitment to social and ecological responsibility. These farmers wanted to make matches that allowed them to see themselves as good stewards of their land, good members of their communities, and good people. Through such matches, small-scale organic farms can remain sites of small-scale resistance to the neoliberalization of agriculture, sites that we hope can be leveraged to lead to larger-scale change. Moreover, we are emboldened by the knowledge that Scenic View—despite its undeniable particularities—is but one of countless small-scale organic farms across the country where farmers do the best they can to forge such matches.

If we allow ourselves to recognize that small-scale agriculture in practice possesses the potential for an alternative to agricultural neoliberalism, there will be risks. As noted, there are affinities among many of the practices advocated by alternative food activists and the political-economic project of neoliberalism. To make such allowances, therefore, is to risk that the efforts pursued by small farmers and food activists to transform the food system could end up furthering its neoliberalization. However, taking such risks is necessary if we are to undermine agriculture's neoliberalization, not at the margins, but through the "transmutation of its core" principles (Bondi and Laurie 2005, 399.).

Risking to Hope in Alternative Agriculture

First, we must risk reconsidering the way we describe consumption. To be clear, we are not endorsing "a political-ecological message of conservation through consumption" (Johnston 2008, 233). Nor are we suggesting a naïve acceptance of the "vote with your dollars" mantra that has taken over much of the alternative food movement. However, as we seek to move the conversation about the potential for alternative agriculture within a neoliberal context forward, we believe that describing consumption solely as an individualistic practice may not be entirely helpful. "Economic processes," like consumption, "should not be set in opposition to extraeconomic cultural and social forces but understood as one special category of social relations" (Zelizer 2010, 367). If consumer relationships to the food system might be reimagined in such a way as to present even a slight challenge to the neoliberal project, we believe it is worth exploring.

It is, in fact, rather characteristic of our neoliberal era to think about consumption as an atomistic process. Such an approach stands in sharp contrast to a long history of conceptualizing consumption as a social, relational process (cf. Veblen 1912; Bourdieu 1984). In stark contrast to a neoliberal view of consumption focused on the individual, scholarship on alternative consumption reveals it to be "eminently social, relational, and active rather than private, atomic, or passive" (Appadurai 1986, 31). Rethinking consumption in this way, as an intrinsically social and collective process, opens the door to imagining a place for consumption in transformative social change. In such a social perspective of consumption, we are no longer obliged to set the presumably individualist concerns of consumption at odds with the collective concerns of politics since "what creates the link between exchange and value is *politics*" (3, emphasis in original). As a result, demand—which drives consumption—is seen as "a function of a variety of social practices and classifications, rather than a mysterious emanation of human needs" or otherwise purely individualistic concerns (29). Looking at consumption from this perspective means looking at consumption as both "sending" and "receiving" social messages—potentially altering economic relationships (31).

There are limits on the ability of consumer demand to manipulate the social and economic forces at work within the agricultural sector. However, moving away from the neoliberal view of consumption, as an individual-level decision, and recognizing its inherently social nature prevents us from foreclosing on potential avenues for building a truly alternative food system. Moreover, recognizing that consumer-producer relationships can be a locus of political action can move us beyond "traditional masculine assumptions of citizenship that overlook how women's 'private' consumption roles contribute to public life" (Johnston 2008, 233). Indeed, women are the most likely purchasers of organic foods, particularly out of concern for the health and safety of their families (Buchler, Smith, and Lawrence 2010; Connolly and Prothero 2008; Lockie et al. 2004). And while consumers cannot simply "vote in" a new agro-economic system every time they check out at the supermarket, the relationships producers and consumers enact together through everyday practices could be the building blocks of a new agricultural economy. At the intersection of so many systems of power, food consumption is inherently political. The question is what politics, and how far can it go?

Second, and perhaps more important, we must also risk allowing for the possibility that the forces of transformative agrarian change will not resemble the progressive politics of the past. While there are limitations to the types of local, decentralized, consumer-based politics endorsed by alternative food activists, we must also recognize the limitations of previous forms of politics. Indeed, the success of progressive politics must be viewed as historically contingent and even anomalous, inextricably linked to the economic and historical context of the Great Depression, World War II, and the subsequent postwar boom. Such massive crises, the associated temporary expansion of the labor movement, and the tremendous engine of economic growth coming out of World War II provided the unique conditions required for the emergence of progressive regulatory politics as we know it (Alperovitz 2011).

However, those same conditions no longer exist and likely will not exist ever again. For one thing, the current scale and global inter-connectivity of government is such that a full-scale economic collapse is unlikely. For another, nuclear capabilities all but preclude the type of industrial-scale war necessary to produce the economic growth seen in the postwar period (Alperovitz 2011). The exceptionalism of the period of progressive change—which we now almost exclusively use as a benchmark for all progressive politics—is such that historians have termed the period beginning with the New Deal reforms "the Long Exception" (Cowie and Salvatore 2008). As a result, while the progressive politics of the past resulted in tremendous social gains, in the absence of the historical, economic, and social preconditions of that era those politics may very well prove ineffectual today.

What does this mean for the future of alternative agriculture? Just as careful ethnographic approaches to farmers lives are methodologically necessary to evaluate farming practices on their own terms without reifying market logics (Pratt 2009), thinking about the possibility of alternative food systems requires the same type of analytic caution. Many individuals are dramatically altering the ways in which they relate to the dominant economy: "They are not merely adopting a private response to what is perforce a collective problem. Rather, they are pioneers of the micro (individual-level) activity that is necessary to create the macro (system-wide) equilibrium, to correct an economy that is badly out of balance" (Schor 2010, 3). Such everyday practices help fill the void caused by the neoliberal retraction of the state: "Localism provides a defensive position. . . . Many people crave face-to-face contact and situations in which they feel they can make a difference" (P. Allen 2004, 169). People have already begun responding to such conditions by changing how they consume, reevaluating wealth, and producing things for themselves or with their communities (Carfagna et al. 2014; Schor 2010; Schor and Thompson 2014).

Agriculture is just one front where producers and consumers are refashioning economic relations and attempting to build alternative economies (Agyeman, McLaren, and Schaefer-Borrego 2013). Tremendous traction has been gained around issues of sharing, for reasons of both ecological sustainability and social inclusion and equity. Much of the attention has fallen on larger, corporate players within the emergent, so-called "sharing economy," specifically on platforms such as Uber and AirBnB. Much like the organic produce that lines the refrigerated cases of our country's chain supermarkets, there may be little "alternative" about such platforms. However, many smaller, nonprofit, grassroots initiatives represent attempts at creating far more alternative market relationships; time banks, where all work is valued equally in terms of the time it takes to do it, have gained newfound popularity, and novel initiatives like food swaps allow individuals to forgo the cash economy and barter at a one-to-one rate of exchange with items they have grown, foraged, or made (Schor and Fitzmaurice 2015). More public-sharing initiatives, rooted in local governments or communities, have also offered new alternatives to reliance on the dominant market for access to resources (Agyeman, McLaren, and Schaefer-Borrego 2013). Changes such as these could produce a more sustainable future not just for agriculture, but also for our broader economy, a model the economist Juliet Schor calls "Plentitude" (2010).

A Plentitude approach to sustainability will require redirecting working hours out of the dominant economy, fostering self-provisioning, valuing experiences of the material world (like the simple pleasures of enjoying a freshly dug carrot), and investing in community. Today, with ever-larger social and ecological challenges to sustainability, we confront a new set of economic conditions shaping how we must pursue alternatives: "[economic] stagnation with low prices, or growth with high costs and mounting damages" (Schor 2010, 20). Many within the dominant economy find themselves working seemingly ever-increasing hours just to keep up. For these overworked Americans, Plentitude offers a way to gain more time and deepen social connections through self-provisioning rather than needing to earn more to meet consumer needs. At the same time, the sluggish growth of the labor market continues to leave countless others unemployed or underemployed. For these underworked Americans, self-provisioning through new economic opportunities within the emergent alternative economy offers critical material and social resources the dominant economy—wracked by both a financial crisis and the increasing ecological pressures of climate change—can no longer sustainably provide.

The food system is a prime location for both observing efforts at building a more sustainable economic system and enacting even more alternative relationships of production and consumption (Schor and Thompson 2014). Small New England farmers like the ones whose stories fill the previous chapters are already making matches among their businesses, their lifestyles, their land, and their consumers that are consonant with the Plentitude model. They represent genuine efforts to create alternative sources of value beyond the economic bottom line and to create new economic relations of production and consumption that vastly differ from those cultivated by the dominant food system. It is our hope that such small-scale resistances on small-scale farms can amount to larger-scale change. However, for this to happen both producers and consumers of the alternative agriculture movement may need to do more to deepen their relationships with each other, with the land, and with their broader communities. In the following sections, we will highlight some of the most salient practical challenges to alternative agriculture in the current context and some suggestions that might contribute to sustainable small-farm success.

A New Agrarian Vision: Reflexive Localism

There are signs in the fields that genuine attempts at alternative agriculture remain possible, even in an era marked by the increasing neoliberalization of agriculture and the conventionalization of organic. As we consider where the opportunities lie for forging alternatives for our time, we will begin by acknowledging the challenges the alternative food movement will undoubtedly face before presenting some suggestions that might possibly contribute to small-farm success. To advance a truly alternative agriculture for our times we must confront the fact that CSA does not automatically produce community around food, that small-scale farms are overwhelmingly operated by producers struggling to keep afloat, that neither social nor environmental sustainability in local food networks can be taken as givens, and, finally, that the efforts of farms and farmers to offer real alternatives in the food system are always at risk of being repackaged in watered-down corporate forms. The success of the farms and farmers described in this book reveals that local, small-scale farms and CSAs can and are working to provide a sustainable alternative food system, but the challenges they face are significant.

First, although we believe small-scale organic farms can build genuine community and politically engage consumers through the use of a CSA model, CSA often falls short in this regard. The CSA model has attributes that are invaluable in creating the type of community that would need to be at the center of a truly alternative food system. There is excellent research showing that participation in CSAs can lead to changes in people's attitudes as they become more involved in alternative relationships of exchange (Thompson and Coskuner-Balli 2007a; Thompson and Press 2014).

Sometimes, the mere structure of the CSA—such as the weekly bag or box of food—is enough to begin more connected forms of consumption as members share what they do not like or offer neighbors, family, or friends their full share when they leave town on vacation. As Thompson and Coskuner-Balli remark, "These forms of sharing may seem inconsequential, but, as a comparison point, consider the likelihood that a consumer would, on a regular basis, purchase excess produce at the grocery store so that these goods could be freely distributed to neighbors and acquaintances" (2007a, 142). More fundamentally, they argue that consumers are, through the process of ideological recruitment, "proactively integrated into a social network linked by a common ideological outlook and goal system and, conversely, that its members develop an enduring sense of commitment toward the community and its core values" (148). In this case, "community" is not based on face-to-face interactions; however, such imagined communities are far from apolitical or inconsequential. Rather, imagined communities are at the heart of national identities, and these consumption communities possess genuine transformative potential.

However, as we saw in our account of Scenic View Farm, although John and Katie hoped to foster community, it often seemed elusive. Apart from the tight community of workers on the farm and the small group of volunteers who helped pack the CSA shares, most members of the CSA picked up their vegetables and exchanged a few pleasantries before going on their way. Likewise, as Laura DeLind reports, research has shown that CSA members are far more likely to have expressed an interest in active participation and engagement in their CSA before their first season, commenting that "reality has a way of bursting well-intentioned, Disney-like bubbles" (2003, 198). Moreover, since CSAs merely demand a seasonal commitment, there is considerably less reason for members to invest effort into the farm commensurate to the effort undertaken by farmers. As a result, De-Lind characterizes member commitments as "casual and highly discretionary" and notes that among thirty-five CSAs surveyed in Michigan, Ohio, and Indiana, twenty-five lacked any working members (ibid). Given such difficulties, we cannot conflate local food with community or assume that CSAs automatically generate the type of community presumed by their name.

These insights are undeniably hopeful; in an agricultural system thriving on economic and not social relations, the ability of CSAs to produce even an ideological commitment to an imagined community is of tremendous consequence. Given the focus on community at the heart of the CSA model, there is reason to hope for a depth of community engagement that is deeper than a sense that "we're all in this together." However, face-to-face communities of engaged producers and consumers must be built, and that takes time and effort. Unfortunately, time is often lacking on small organic farms operated by overworked farmers, who are already contributing more than their fair share of effort into transforming America's food system.

The lack of additional time on small farms due to the tremendous amount of work involved in their operation brings us to the second challenge facing an alternative food movement committed to genuine alternative agriculture. Although programs like CSA have helped many small farmers survive—as we observed on Scenic View and the other small farms we discussed—many small farmers are still struggling, including John and Katie. While direct-to-consumer sales have helped keep these farms in business, they still aren't thriving. As we discussed in the previous chapter, while farmers like John, Katie, and the others in our study have found ways to create sustainable businesses, they haven't found a way to enjoy the personal security and social stability at the heart of the American Dream. If the alternative agriculture movement is to be truly alternative, creative solutions are necessary to help ensure that small-scale farming pays. It should not be so hard for the countless hard-working individuals like John and Katie to farm in ways that are environmentally, economically, and socially sustainable.

A third challenge to alternative agriculture in the food system involves the issue of justice. While the localization of food production has tremendous potential to address issues of justice by bringing the inequalities of production and consumption into view, the local food movement has often been guilty of ignoring issues of justice and equity (P. Allen 2004; Harrison 2011; P. Allen and Sachs 1993). Even when alternative food movements address issues of justice, they tend to revolve around problems of access and urban food security, not the conditions of farmworker employment. And as long as organic remains "just another method of agriculture" (Warner 2006, 2), there will remain profound issues of injustice for farmworkers unable to find employment on pesticide-free farms (see also Harrison 2011; P. Allen 1993). However, research has shown that American consumers have expressed a willingness to pay a premium for food with the assurance that farmworkers were paid a living wage and labored under safe working conditions, with frequent organic consumers willing to pay even greater premiums (Howard and Allen 2008).

While localizing food production has great potential to make issues of justice visible, there are also risks to those left out of these emerging solutions. CSA programs often advertise their environmental sustainability measures through organic certification or descriptions of pesticide-free, beyond organic farming techniques, but how often are socially sustainable labor conditions advertised? As we saw with Scenic View Farm, many small farms can provide good jobs with safe—even enjoyable—working conditions. However, until issues of justice become central to the alternative food movement, a socially just alternative is far from assured.

Despite the frequent claims about environmental sustainability, even on this issue the alternative food movement faces challenges. Sustainability claims of the local food movement often rest upon "food miles" arguments. For example, shipping a head of iceberg lettuce from California to the East Coast uses thirty-six times the calories in fossil fuel than the lettuce could provide (McKibben 2007, 65). Such shipping represents a genuine waste of energy, and local food systems could undoubtedly reduce such waste. However, when it comes to food sustainability, economies of scale may matter, and driving out to the farm (particularly from urban or suburban areas) to pick up a share of vegetables may not be the most energy efficient approach. In fact, Matthew Mariola argues, "From the point of view of sustainable energy use, farm-direct sales present a catch-22. On the one hand, the food has not traveled more than a mile from where it was produced to where it is purchased, in comparison with the 20–50 miles from a farm to a farmers' market, let alone the thousands of miles from California. On the other hand, the consumer must engage in the highly energy-intensive act of traveling to a farm to buy only a handful of products . . . [that] account for far more energy per item when purchased on-farm than when purchased in a grocery store as part of a whole shopping list" (2008, 195). In a dramatic reversal of the typical "food miles" logic, research has shown that for some products, such as apples, lamb, and milk, it is less carbon-intensive for them to be produced in New Zealand and shipped to the UK than to have them produced domestically (Saunders, Barber, and

Taylor 2006; see also Hess 2009). This is not to say that local food systems could not be fashioned to be more environmentally sustainable. Rather, it is safe to say that the fact that CSAs are highly reliant on the oil economy is reason enough to work on their improvement.

Finally, as the case of Scenic View Farm demonstrates—along with the numerous farms like it across the United States—small-scale, localized systems of organic production can represent genuine and potentially effective alternatives to the business-as-usual approach of agribusiness food production. The history of the organic food movement reveals that there is always the threat of co-optation. We began this book by providing an account of how organic not only expanded, but also profoundly changed over the course of its more than seventy-five-year march from the agricultural margins to supermarket shelves in Whole Foods, Walmart, and the like.

With the entry of conventional retail giants like Walmart into the business of organic, Whole Foods began highlighting products from local farmers (Ness 2006). Now, even Walmart has instituted a program aimed at buying from small farmers. Walmart's new Heritage Agriculture program allows farmers within a day's drive of a distribution center to bring their produce to be purchased by the company (Kummer 2010). As of 2010, only 4–6 percent of Walmart's food was coming from farmers participating in the program; however, Walmart hoped to increase the share to 20 percent of its total produce (ibid.). Although we reported above that the move toward "local" food as the paragon of responsible eating was seen as a response to the supermarketization of organic food, it would appear that local food is hardly immune to criticism. Rather, to echo our appeal above to the work of Herbert Marcuse, "Liquidation of *two-dimensional* culture takes place not through the denial and rejection of the 'cultural values,' but through their wholesale incorporation into the established order, through their reproduction and display on a massive scale" (1964, 57). The ability of the alternative agriculture movement to remain a viable "alternative" to industrial farming may be most threatened by the reproduction of its message and least threatening practices by the conventional food system—without the commitment toward improving agricultural sustainability demonstrated by individuals like Katie and John at Scenic View.

The trouble, then, as we envision alternative solutions for the contemporary food system, is that alternative agriculture needs to be a moving target in order to keep its "alternative" edge. Conventional agribusinesses and the agro-industrial organic complex are constantly responding to the changes pioneered by the alternative food system, incorporating the symbols of alternative and displaying them on a massive scale. As a result, the inside of Whole Foods increasingly resembles a farmers' market: it now even captures the smiling faces of farmers on signs beside the produce. And we should not expect that conventional retailers like Walmart will be far behind. This phenomenon of co-optation is a challenge for the alternative food movement, as its symbols of what constitutes alternatives to the mainstream are rendered meaningless through their increasing alignment with the agricultural system they purport to challenge. However, the challenge of corporate co-optation can compel us to push toward *even more alternative* solutions. By creatively pushing the boundaries of the possible in search of deeper social, environmental, and economic sustainability in contemporary agriculture—keeping what it means to be organic a moving target—local farms and local communities can, and are, working to create an alternative food system.

Vanguards of Agrarian Change: Organic as a Moving Target

Alternative agriculture faces challenges on farms like Scenic View across the country. Nevertheless, we believe that the tendency toward co-optation can present the vanguard of the alternative food movement a tremendous opportunity. If agribusiness and industrial-organic are a moving target, the alternative food movement must be too. As Michael Bell has argued, sustainable agriculture isn't just about the amelioration of the worst tendencies of the industrial food system; it must be about "new cultivations of farming. And that means both new senses of what a farm is and can be and new senses of what a person is and can be" (2004, 203). As small-scale local farmers innovate and the corporate food system appropriates those innovations, we push the entire food system forward. However, to be effective, alternative agriculture must continue to push at the margins of the food system, ensuring that "alternative" remains meaningful.

Small-scale farmers like those highlighted in this book—who confront the challenges to being alternative in their daily lives and practices—must set the course for making agriculture more sustainable. These farmers already demonstrate the possibility of truly alternative agriculture in our time. However, the challenges they face and their struggles to remain viable reveal the need for deepened support and investment in their efforts. As a result, we need to focus on building increasingly alternative, reflexive local food systems. In light of these challenges, here are our suggestions.

First, CSA can be a valuable tool for connecting consumers to local food, but it often falls short of fundamentally transforming producer/consumer relationships, so pushing this alternative agricultural tool forward requires that we open the CSA "black box." In other words, we need to think of the CSA as only a starting point for alternative agriculture, not the end point, and allow ourselves to rethink its configuration. Most farmers think about CSAs as a business strategy for survival, while most consumers view them as a food-buying strategy (albeit a novel value-added one). Unfortunately, these two understandings of CSA do not leave much room for *community*. Both farmers and consumers could consider refashioning the CSA model in a way that prioritizes sustainable communities, making it a Community Supported *Community* Agriculture model. As Michael Bell suggests, "The image of the farm tends to guide our thinking back to a sense of agriculture as the beyond we left behind." Instead, much like John and Katie, he offers the image of the "garden" as a truly alternative vision of agriculture, contending, "Cultivation . . . is a task not only for those people we have long regarded as the agriculturalists. It is a task not only for farmers. It is a task for everyone" (2004, 248). Research on community gardens has shown that such efforts can serve as a locus of civic activism that can help expand public space, improve the quality of life in local environments, and enhance local food security (Nettle 2014). Rather than relying on a small number of farmers to support our communities, more communities could collaborate to support themselves, of course with farmer involvement, but perhaps in a nuanced and expansionary way.

Opening the CSA black box and evaluating new directions for the CSA model could even force us to rethink what counts as a farm. Community Supported Community Agriculture could take place in centralized gardens or farms could be formed out of a patchwork of front lawns, urban rooftops, and even balconies. Either way, with community at the heart of such an approach, farms could act as key components to an expanded network of reflexive localism. Weekly or biweekly neighborhood gatherings (complete with music and beverages suitable for both children and adults, of course) could be used to accomplish farm tasks and build the type of face-to-face community so often missing from the current CSA model. Working together, communities could even host fall canning parties to help stop CSAs from existing as a mere seasonal commitment. Whatever the approach, to build meaningful communities and to transform producer/consumer relationships around food, the local food movement might be well served by a new focus on increasing the number of people producing their own local food—not just consuming it.

Many consumers, just like John, Katie, and the other workers with whom we spoke, are also overworked. We have already suggested concerns that the demands of building alternative food systems are unreasonable for such consumers and can disproportionately fall upon women—hardly an alternative set of social relations (Johnston and Szabo 2011; Sandilands 1993). Nonetheless, what we are suggesting is not that labor exploitation be outsourced from small-scale farmers to the consumers of alternative agricultural commodities. CSA programs and other local food initiatives already represent economic matches that tremendously reduce the distance between producer and consumer from what is experienced on a hectic weeknight in the supermarket produce aisles. However, incorporating the concepts of

Plentitude could deepen—and fundamentally transform—the disjuncture between production and consumption that the exchange relations of many alternative food initiatives continue to reproduce.

Commitments to participate in the production of truly sustainable food would represent a reallocation of time for many overemployed Americans that would deepen opportunities for social connections, provide meaningful attachments to the environment and the food system, and replace cash earnings with no less tangible material resources. Moreover, for those struggling for work in the dominant economy, establishing agricultural systems with opportunities for productive work could prove a critically important means of food sovereignty. At Scenic View, one of the volunteers who packed the CSA shares was an underemployed mother. Not only did she enjoy getting out of the house and connecting with the others who volunteered, but she also was able to leave each week with a bag of fresh produce in exchange for her help.

While many scholars have worried that consumer citizenship places the burden of tremendous social and ecological problems within the food system on ordinary shoppers, others have highlighted the ways many regions struggling with economic insecurity have often worked to develop alternatives to the dominant unsustainable economy. These systems are predicated on reciprocal exchange, communal production, and self-provisioning (Gowan and Slocum 2014; McEntee 2011). As a result, "Emphasizing the economic benefit to the individual . . . of *growing, producing, or bartering* locally in order to save money would be more effective in garnering support for accessible and therefore socially just food systems (McEntee 2011, 253). In an increasingly global food system in which agricultural prices set in distant markets have profound local consequences, community investment in food production could help regain local control of and access to food resources (W. Allen 2013). Certainly involvement in this type of non-market voluntary work would inevitably mean very different things for those more or less privileged by the broader economy. However, establishing exchange relationships around food that are even more socially connected and reciprocal could provide greater sustainability, for numerous reasons, for many in the food system.

Second, while CSAs have helped many small farmers survive, across the country farmers like those we encountered at Scenic View are still struggling. While farm-to-school programs have become all the rage, perhaps we need to start thinking more seriously about the opposite, about school-to-farm initiatives. Colleges and universities are repositories of capital: human, social, and economic. Meanwhile, farmers struggle to make sustainable incomes in light of costs—which include labor. The alternative food movement could work to forge public or quasi-public partnerships between schools and farms. Colleges and universities could provide interns to sustainable farms—providing labor for credit—helping to reduce production costs while giving students practical insights into the difficulties associated with true sustainability. Moreover, colleges are large consumers of farm products. Rather than merely highlighting local produce in dining halls, perhaps university endowments could be leveraged to *invest in the community*—helping to build more robust food systems with more targeted support than dubiously "voting" with their dollars.

Third, it is critical for alternative food systems to address issues of justice if they are to become truly alternative. Here, reflexivity could be especially helpful, but it cannot be limited to the checkout line. Rather, a critical engagement with issues of justice is necessary across all phases of the food system. Consumers within the movement must bring a concern for farm working conditions to bear on their purchasing decisions, in addition to the more typical concerns of environmental sustainability and quality. Farmers must also take issues of justice seriously. This can be difficult, especially for small farmers, given the fact that "the chronic economic insecurity at the heart of their industry" often means that "the profits of most small farms today are directly predicated on workers' low wages" (Gray 2013, 149). However, many local farmers do desire socially and economically sustainable employment practices for their workers—despite the costs. In Margaret Gray's work, *Labor and the Locavore* (2013), several of the farmers with whom she spoke in the Hudson Valley described the need for better labor practices in the local food system. We have already seen that the small-scale organic farmers at Scenic View treated their workers well, such that the employees enjoyed their time on the farm, even hanging around after work. Far from being exploited, Joey remarked that he often felt as though he was exploiting the kindness and generosity of John and Katie. However, these behind-the-scenes labor practices were not part of what consumers were purchasing.

The reality is that many farmers within the sustainable agriculture movement—like their conventional counterparts—often have to exploit their own labor to keep farm businesses afloat (Pilgeram 2011), an experience echoed in the lives of Katie and John at Scenic View Farm. And many often fall into the same unsustainable labor practices as their conventional counterparts in order to remain profitable. However, both forms of exploitation tend to be minimized by small-farm supporters through appeals to the romanticized history of family farming (Gray 2013). To this end, the alternative food movement should increasingly advocate for food labels like "organic" to reflect the social and ecological conditions of production, and alternative farmers' need to educate consumers about their own labor practices.

Several organizations, such as the Agricultural Justice Project, the Domestic Fair Trade Association, and the Equitable Food Initiative, provide certifications for producers who protect the rights of food system workers. And many American consumers already express interest in labeling schemes that address social and economic concerns beyond the purview of the organic label and a willingness to pay a premium to ensure that American farmers and farmworkers work in conditions that are socially and economically sustainable—not just pesticide free (Howard and Allen 2008, 2010). Such concerns have already been articulated in the fair trade label, certifying a level of social sustainability for farmers and farmworkers in the global south. If organic agricultural practices are to represent a true alternative to a failed system of industrial food production, socially unsustainable practices must be redressed—both at home and abroad. While "certified organic is a production standard in a codified system of 'allowable' substances in the field . . . if stakeholders could be convinced of the efficacy of Fair Trade as a unifying principle for [Alternative Food Networks], it could provide verification of the social sustainability industrial organic does not provide" (Duram and Mead 2013).

Beyond such efforts, reflexivity about issues of justice needs to occur at the level of community. The types of economic relations forged in local food systems are already designed to promote a more socially connected food system than industrial-scale agriculture can provide. These types of connections are currently used to connect people with the material qualities of their food, such as its seasonality, and to connect food to the businesses that provide it. However, these more relational forms of production and consumption could just as easily be used to connect people to the labor practices of local farms so that the sustainable and just treatment of farmworkers becomes part of what supporters of local food producers buy into. From the perspective of both sustainability and the viability of small-farm businesses, Gray remarks, "Think about the benefits to be accrued from restructuring agriculture around sustainable jobs and how improved labor practices can be promoted as a selling point. Just as Whole Foods has a point system to indicate how the animals that become our meat were treated, so, too, smaller farmers could advertise the benefits they offer their workers to explain food costs to consumers" (2013, 149). We are hopeful that the types of matches small-scale local farmers are already making with their consumers could easily be deepened to make such connections a reality.

Labor protections could also be facilitated by the previous two suggestions about Community Supported Community Agriculture programs and institutional partnerships with the food system. Community Supported Community Agriculture could be developed with clear social justice mandates, while public and quasi-public partnerships between college students and farms might help bring institutional forces to bear on establishing safe and fair farmworker conditions. This type of deep sustainability—recognizing the need for human relationships "that ensure fairness with regard to the common environment and life opportunities" (IFOAM 2013)—was at the heart of the historic organic movement. Such concerns must be addressed in new ways for our time by contemporary participants in the organic food system if we are to ensure that the work of farmers like John and Katie is socially sustainable and secure—not just free of toxic pesticides and synthetic fertilizers. Farmers cannot continue to shoulder the burden—pressured to exploit their own labor or that of others—to remain viable businesses in their pursuit of alternatives to the industrial agricultural system.

Fourth, to be truly alternative, alternative food systems must address the challenges of ecological sustainability in a more powerful way. Local food systems can solve the problem of food miles; however, if we look at all of the food miles involved in local food systems, we see there are tremendous inefficiencies. Once again, solving the problems that limit the scope of alternative agriculture requires opening the black boxes of CSAs, farmers' markets, and farm stands and examining the role of communities/consumers. All of these can produce excess miles, especially if we consider urban residents driving to distant farms. The problems are not insurmountable. One solution is to situate CSA drop-off points at conventional food retailers. Most people do not live entirely off of their CSA produce, so driving out to the farm to pick up their share represents an additional fossil-fuel-dependent trip for food. With the distribution of CSA shares centrally located at a standard supermarket, customers could one-stop shop while still supporting their CSA farm. Alternatively, farms could require that suburban customers receive their CSA shares as a delivery so that one truck could be used to drop off all of the shares using the most energy-efficient route. It would not be hard for a savvy college student to develop a "Green CSA pickup" phone app to map out and identify energy-efficient CSA routes.

Solving the energy problems of local food systems will also require state support, at both the federal and municipal levels. To build an ecologically sound food system, it will be critical to gain federal support through initiatives to equip small-scale farms with the infrastructure to be more efficient and the land grant research funds to identify and create more competent local food system models. At the municipal level, cities serious about supporting more sustainable food systems could invest in the old concept of public markets. Such markets could more efficiently bring CSA shares and other farm products to urban residents while providing a more democratically accountable food space than the conventional supermarket.

The use of energy and other inputs *on* local farms, not just in food transportation, will also require careful, often creative, solutions if our food system is going to become genuinely environmentally sustainable. At the headquarters of urban farming pioneer Will Allen's organization, Growing Power, in Milwaukee, nearly closed systems have been created to raise both fish and vegetables, reducing the need for inputs. The waste generated by farmed tilapia is filtered through beds of tomatoes, strawberries, and watercress, fertilizing the plants and filtering the water. Approximately seventy-five volunteers, with no fossil-fuel-burning equipment, built the system, which required a three-foot-deep trough dug the entire length of the greenhouse. In the winter, they are able to grow spinach in a greenhouse shared by chickens—the heat generated by the birds is just enough to keep the greenhouse warm. The farm also makes use of anaerobic digesters to generate methane from its food waste, which it can then burn to produce heat or electric energy. Allen thinks of food and farming waste as an untapped resource since the United States produces enough waste to power one in twenty American homes (W. Allen 2013). Recall that at Scenic View, Katie wanted electric golf carts to get workers and equipment across the farm. For the time being, though, it was an aspiration. Between the tractors and the pickup trucks, the smell of burning diesel fuel (not hay or fresh-cut grass) lingered on clothing many days after work in the fields. While not every sustainable farm will develop aquaponic systems for producing food and fertilizer or heat greenhouses with anaerobic digesters, ecological sustainability will likely require such types of investments in smart energy solutions.

Some of the inefficiencies of emergent local food systems are even more fundamental. They are built into the ways we think about food and the ways we eat. As we saw with John and Katie, decisions about what to plant often centered on what consumers wanted; consumers wanted variety, they wanted tomatoes, and they wanted replacement crops to fill in when a favorite item in their CSA share didn't come through. And by and large, the farm provided what consumers wanted. Dan Barber, one of the leading chefs behind the farm-to-table movement, has remarked that there is something fundamentally unsustainable about the way we currently conceptualize local food systems: "Farm-to-table chefs may claim to base their cooking on whatever the farmers pick that day (and I should know, since I do it often), but whatever the farmer has picked that day is really about an expectation of what will be purchased that day. . . . It forces farmers into growing crops like zucchini and tomatoes (requiring lots of real estate and soil nutrients) or into raising enough lambs to sell mostly just the chops, because if they don't, the chef, or even the enlightened shopper, will simply buy from another farmer" (2014, 15). Indeed, only once did we observe environmental efficiency shape what crops consumers received: a local sleep-away summer camp received komatsuna—an Asian green that produces leaves bigger than collards—because it merely requested something green and nutritious to feed the campers, and the farm could produce it with very little space or added inputs. For local food systems to become more environmentally sustainable, it will take major changes in the ways farmers, chefs, and everyday consumers think about food and not just in terms of from how far away it should come.

Finally, alternative food movements must work to keep their alternative character a moving target in order to ensure that "alternative" doesn't become merely a watered-down niche market within the agribusiness model. As we push the alternative model further, it will be co-opted, and it will be watered down, but we can push the broader conversation forward while forging a deeper level of sustainability. As we have seen, small farms like Scenic View have done a remarkable job remaining faithful to the environmental beliefs that motivated the organic movement, despite trends toward greater conventionalization within organic. The neoliberal economic paradigm has done much damage to the movement; however, the opportunities for constantly recreating alternative food systems are not foreclosed.

To forge an alternative agriculture for our time—to ensure community, justice, and ecological sustainability—small farmers like John and Katie will need to push the boundaries of what it means to be "alternative." But they cannot do it alone. Small farms are struggling to stay afloat, let alone create ever-new forms of alternatives to the conventional food system. The organic of today is not that of the past, and that of tomorrow cannot be what it is today. To succeed, we cannot look to the movement of the past for easy answers. Rather, we will need to forge new, collaborative solutions to the challenges small farms face; we will need to grow a new agrarian vision rooted in community, not rugged individualism. If small farms need to push the boundaries of the agricultural system, they will require communities to help push those boundaries with them.

APPENDIX: METHOD AND APPROACH

To discern how emotions, morals, and personally meaningful so-cial relationships shape how organic farmers do their jobs and what being an organic farmer means to them, our working alongside them in the fields was critical. Excellent studies of the relational under-pinnings of markets have been conducted using ethnographic meth-ods and without the authors having spent substantive time in the field (e.g., Healy 2006). However, we were interested in the everyday practices, understandings, and interpersonal exchanges that both shaped and gave meaning to the work of organic farmers but that could just as easily be taken for granted. Consequently, our research follows in the tradition of participant observation, subjecting oneself to the various circumstances and everyday realities that shape our research participants' "response to the situation" (Goffman 1989, 125). Our study also reflects new developments in ethnography that stress the importance of "going along" with our research subjects since how people understand the world is based on their place in it—both physically and metaphorically (Kusenbach 2003). To achieve this perspective and insight, Connor Fitzmaurice worked on Scenic View Farm as an intern.

In many ways, taking an ethnographic approach to the study of an organic farm was easier said than done. Fitzmaurice undertook the first stage of fieldwork for this book in 2009. We contacted a dozen organic farms that were both small (all of the farms were under ten acres) and that reported utilizing organic practices (whether certi-fied or not). We requested that Fitzmaurice be permitted to work full time at no cost in exchange for access to the farm and its associ-ated members (hired hands, CSA members, visitors to the farm, etc.). Nine of the farms we contacted never responded. Two expressed reservations and eventually rejected our request, one because it was no longer certified organic and the other because it felt uncomfortable having an unpaid laborer. But one farm, Scenic View, at least entertained the idea at the outset. However, the owners, John and Katie, suggested that if Fitzmaurice was going to "help out" around the farm while conducting interviews, they wanted to "try him out to see if he's actually going to be helpful."

Undoubtedly, part of the problem of access was that in many ways, Fitzmaurice was "studying up." As we have highlighted, the farmers at Scenic View—like many farmers in New England—are quite privileged along the lines of race, class background, and cultural capital, at least in comparison to farmers and farmworkers from around the country. Sociologists in general, and ethnographers in particular, have a long history of studying the marginalized of society. However, a new wave of scholarship has pointed to the importance of studying those with economic, social, and cultural power in order to understand how privileged social worlds are produced and maintained. The fields of an organic farm are seemingly far less glamorous than the social worlds at the center of many of these path-breaking studies (Ho 2009; Khan 2012; Mears 2011). Farmers, however, have come to receive heroic status in the eyes of many ethical consumers. For example, they are clamored after by celebrity chefs and included in alternative food discourses that privilege the white tradition of yeoman farming (Alkon and McCullen 2011; Carfagna et al. 2014; Guthman 2011). Consequently, Fitzmaurice's access to Scenic View was not guaranteed, and it should not be taken for granted that it was easily achieved.

Gaining access to a field site was made easier by Fitzmaurice's own privilege. As a white, college-educated young adult with a background in environmental studies, he was accepted as a suitable worker by John, Katie, and their employees after a day working in their fields. Over the course of the 2009 growing season, he took on the role of "observing participant" rather than participant observer (Wacquant 2004, 6). This description means that he lived on the farm and often worked ten-hour days in the field, engaging in all of the tasks involved in an organic farm. This also meant learning to hoe, to decide what beets were ready to be pulled based on the size of their leaves, to harvest heads of lettuce with a carefully wielded blade, and the like. As a result of the largely unmechanized labor practices of Scenic View Farm, workers were often in close proximity to each other, engaged in repetitive tasks—allowing Fitzmaurice to conduct informal interviews over the course of the workday. In addition, he also conducted semi-structured interviews, lasting from one to three hours, with all seven of the individuals working at Scenic View that season. After the 2009 growing season, informal follow-up interviews were conducted with John and Katie during 2010 and 2011.

In addition to our ethnographic case study of Scenic View, we conducted five months of participant observation at seven farmers' markets during the spring and summer of 2012 in a large northeastern city and in its suburbs in 2012–2013. The purpose of these additional observations was to investigate how New England farmers described their practices during routine exchanges with their customers. A total of fifteen semi-structured interviews were also conducted with New England farmers and farmworkers. These farmers were recruited at the farmers' markets, with interviews lasting approximately one hour or less. Through these interviews, we explored the web of social connections, moral obligations, and emotional commitments shaping different experiences of "organic" across a spectrum of farms, ranging from certified to conventional. In total, this research is based upon over four hundred hours of observing the relationships and asking about the social, moral, and emotional exchanges of contemporary farmers.

In order to understand the relational work of organic farmers' practices and economic exchanges, it was equally important to situate our observations in the shifting and historically contingent meanings of the term "organic." Good matches are often achieved through appeals to familiar, culturally available scripts and are dependent on shared systems of evaluation (Wherry 2012). Given our theoretical framework, we have chosen to utilize primary source documents from various historical periods of the organic movement. These serve to demonstrate the types of matches that not only gave rise to the organic movements of the past, but also shape the meanings contemporary organic farmers attach to their own economic lives. To do so, we compiled news reports from major newspapers (many published by the Associated Press), as well as recorded speeches, minutes from governmental committee hearings, government investigative reports, and regulatory documents from each significant period in our historical account of the organic movement. Statistical data were obtained from governmental and non-governmental sources such as the USDA, the USDA Agricultural Marketing Service, the USDA National Agricultural Statistics Service, the Organic Trade Association, the University of Massachusetts Extension, and the International Federation of Organic Agriculture (IFOAM). By drawing from these primary sources, we are able to show not only the historical development of the organic movement, but also some of the important social, cultural, and moral forces that have shaped the development of the organic market.

Finally, we chose to draw on many excellent existing accounts of the practices of contemporary organic farmers. The careful work of scholars such as Leslie Duram, Brian Donahue, Michael Bell, Patricia Allen, Alison Hope Alkon, and Brad Weiss (to name just a few of those found on these pages) provides rich descriptions of how contemporary farmers make alternative approaches to agriculture work in practice. By drawing comparisons between our findings and those of these studies, we are able to theorize on the important role of the relational work occurring on today's alternative farms: shaping and giving meaning to "organic" in the lives of today's alternative farmers.

REFERENCES

Adams, D. C., and M. J. Salois. 2010. "Local versus Organic: A Turn in Consumer Preferences and Willingness-to-Pay." *Renewable Agriculture and Food Systems* 25, no. 4:331–341.

Agricultural Marketing Service. 2008. "Organic Labeling and Marketing Information." U.S. Department of Agriculture. http://www.ams.usda.gov/AMSv1.0/getfile?dDocName=STELDEV3004446. Accessed July 22, 2012.

———. 2009. "Facts on Direct-to-Consumer Marketing." U.S. Department of Agriculture. http://www.ams.usda.gov/AMSv1.0/getfile?dDocName=STELPRDC5076729. Accessed July 19, 2012.

———. 2011. "Farmers Market Growth: 1994–2011." U.S. Department of Agriculture. http://www.ams.usda.gov/AMSv1.0/ams.fetchTemplateData.do?template=TemplateS&leftNav=WholesaleandFarmersMarkets&page=WFMFarmersMarketGrowth&description=Farmers%20Market%20Growth&acct=frmrdirmkt. Accessed July 19, 2012.

Agyeman, J., and B. Evans. 2004. "'Just Sustainability': The Emerging Discourse of Environmental Justice in Britain?" *Geographical Journal* 170, no. 2:155–164.

Agyeman, J., D. McLaren, and A. Schaefer-Borrego. 2013. "Sharing Cities." Friends of the Earth briefing paper.

Alasia, A., A. Weersink, R. D. Bollman, and J. Cranfield. 2009. "Off Farm Labour Decisions of Canadian Farm Operators: Urbanization Effects and Rural Labour Market Linkages." *Journal of Rural Studies* 25, no. 1: 12–24.

Alkon, A. H. 2008. "Paradise or Pavement: The Social Constructions of the Environment in Two Urban Farmers' Markets and Their Implications for Environmental Justice and Sustainability." *Local Environment* 13, no. 3:271–289.

Alkon, A. H., and C. G. McCullen. 2011. "Whiteness and Farmers Markets: Performances, Perpetuations . . . Contestations?" *Antipode* 43:937–959.

Allen, P., ed. 1993. *Food for the Future: Conditions and Contradictions of Sustainability.* New York: John Wiley and Sons.

———. 2004. *Together at the Table: Sustainability and Sustenance in the American Agrifood System*. University Park, PA: Pennsylvania State University Press.

Allen, P., and M. Kovach. 2000. "The Capitalist Composition of Organic: The Potential of Markets in Fulfilling the Promise of Organic Agriculture." *Agriculture and Human Values* 17:221–232.

Allen, P., and C. Sachs. 1993. "Sustainable Agriculture in the United States: Engagements, Silences, and Possibilities for Transformation." In P. Allen, *Food for the Future*.

Allen, W. 2013. *The Good Food Revolution: Growing Healthy Food, People, and Communities*. New York: Penguin.

Alperovitz, G. 2011. "The Emerging Paradoxical Possibility of a Democratic Economy." Plenary keynote address, Association of Social Economics, Denver, CO.

Altieri, M. A. 1995. *Agroecology: The Science of Sustainable Agriculture*, 2nd ed. Boulder, CO: Westview Press.

Appadurai, A. 1986. *The Social Life of Things: Commodities in Cultural Perspective*. Cambridge: Cambridge University Press.

Arcury, T. A., and S. A. Quandt. 2009. *Latino Farmworkers in the Eastern United States: Health, Safety and Justice*. New York: Springer.

Associated Press. 1981. "30 States Get Federal Warning on Tainted Cans of Mushrooms." *New York Times*, May 23.

———. 1982. "Contaminated Milk Blamed for Infection in 172 Southerners." *New York Times*, September 25.

———. 1984. "Residues on Food Cause Much Concern." *New York Times*, March 28.

———. 1989a. "Health Official Rebukes Schools over Apple Bans." *New York Times*, March 16.

———. 1989b. "Organic Produce Preferred." *New York Times*, March 21.

———. 1989c. "Government Will Buy Apples Left Over from Scare on Alar." *New York Times*, July 8.

Balfour, Lady E. 1977. "Towards a Sustainable Agriculture—The Living Soil." *Proceedings of IFOAM International Organic Farming Conference*. http://soilandhealth.org/wp-content/uploads/01aglibrary/010116Balfourspeech.html. Accessed December 15, 2015.

Bandelj, Nina. 2012. "Relational Work and Economic Sociology." *Politics and Society* 40, no. 2:175–201.

Barber, D. 2009. "You Say Tomato, I Say Agricultural Disaster." *New York Times*, August 8. http://www.nytimes.com/2009/08/09/opinion/09barber.html?pagewanted=all. Accessed July 19, 2012.

———. 2014. *The Third Plate: Field Notes on the Future of Food.* New York: Penguin.

Beavan, C. 2009. *No Impact Man: The Adventures of a Guilty Liberal.* New York: Farrar, Strauss, and Giroux.

Beecher, N. A., R. J. Johnson, J. R. Brandle, R. M. Case, and L. J. Young. 2002. "Agroecology of Birds in Organic and Nonorganic Farmland." *Conservation Biology* 16:1620–1631.

Beeman, R. S., and J. A. Pritchard. 2001. *A Green and Permanent Land: Ecology and Agriculture in the Twentieth Century.* Lawrence, KS: University Press of Kansas.

Belasco, W. J. 2007. *Appetite for Change: How the Counterculture Took on the Food Industry.* Ithaca, NY: Cornell University Press.

Bell, M. M. 1989. "Did New England Go Downhill?" *Geographical Review* 79, no. 4:450–466.

———. 2004. *Farming for Us All: Practical Agriculture and the Cultivation of Sustainability.* University Park, PA: Pennsylvania State University Press.

Berger, P. L., and T. Luckmann. 1966. *The Social Construction of Reality: A Treatise in the Sociology of Knowledge.* New York: Doubleday.

Berlin, L., W. Lockeretz, and R. Bell. 2009. "Purchasing Foods Produced on Organic, Small, and Local Farms: A Mixed-Method Analysis of New England Consumers." *Renewable Agriculture and Food Systems* 24, no. 4:267–275.

Berry, W. 1990. "Nature as Measure." In W. Berry, *What Are People For?* New York: North Point Press.

———. 2002. *The Art of the Commonplace.* Washington, D.C.: Counterpoint.

Bhatnagar, P. 2006. "Wal-Mart's Next Conquest: Organics." *CNN Money Magazine.* http://money.cnn.com/2006/05/01/news/companies/walmart _organics/. Accessed July 19, 2012.

Blanding, M. 2002. "The Invisible Harvest." *Boston Magazine*, October. http://www.bostonmagazine.com/2006/05/the-invisible-harvest/. Accessed July 13, 2015.

Blank, S. C. 1999. "The End of the American Farm." *The Futurist*, April, pp. 33–36.

Blatt, H. 2008. *America's Food: What You Don't Know about What You Eat.* Cambridge, MA: MIT Press.

Block, D. R., N. Chavez, E. Allen, and D. Ramirez. 2011. "Food Sovereignty, Urban Food Access, and Food Activism: Contemplating the Connections through Examples from Chicago." *Agriculture and Human Values* 29, no. 2: 203–215.

Bondi, L., and N. Laurie. 2005. "Introduction." *Working the Spaces of Neoliberalism: Activism, Professionalism, and Incorporation.* Oxford: Blackwell.

Born, B., and M. Purcell. 2006. "Avoiding the Local Trap: Scale and Food Systems in Planning Research." *Journal of Planning Education and Research* 26:195–207.

Bourdieu, P. 1977. *Outline of a Theory of Practice.* Cambridge: Cambridge University Press.

———. 1984. *Distinction: A Social Critique of the Judgment of Taste.* Cambridge, MA: Harvard University Press.

Brady, D. 2006. "The Organic Myth." *Bloomberg Businessweek.* http://www.businessweek.com/stories/2006-10-15/the-organic-myth. Accessed July 19, 2012.

Brehm, J. M., and B. W. Eisenhauer. 2008. "Motivations for Participating in Community Supported Agriculture and Their Relationship with Community Attachment and Social Capital." *Southern Rural Sociology* 23, no. 1:94–115.

Brenner, N., and N. Theodore. 2002. "Preface: From the 'New Localism' to the Spaces of Neoliberalism." *Antipode* 34, no. 3:341–347.

Brown, A. 2001. "Counting Farmers Markets." *Geographical Review* 91, no. 4:655–674.

Buchler, S., K. Smith, and G. Lawrence. 2010. "Food Risks, Old and New: Demographic Characteristics and Perceptions of Food Additives, Regulation and Contamination in Australia." *Journal of Sociology* 46:353–374.

Buck, D., C. Getz, and J. Guthman. 1997. "From Farm to Table: The Organic Vegetable Commodity Chain of Northern California." *Sociologia Ruralis* 37, no. 1:3–20.

Busch, Lawrence, Jeffery Burkhardt, and William B. Lacy. 1992. *Plants, Power, and Profit: Social, Economic, and Ethical Consequences of the New Biotechnologies.* Oxford: Blackwell.

Buttel, F. 2001. "Some Reflections on Late Twentieth Century Agrarian Political Economy." *Sociologia Ruralis* 41, no. 2:165–181.

———. 2006. "Sustaining the Unsustainable: Agro-food Systems and Environment in the Modern World." In *The Handbook of Rural Studies*, ed. P. Cloke, T. Marsden, and P. Mooney, 213–230. London: Sage Publications.

Cagle, J. 2011. "Food, Farm, Family." *Register Guard*, May 25. http://www.registerguard.com/web/specialtastings/26018789-47/farm-csa-deck-family-eggs.html.csp. Accessed July 19, 2012.

Caldwell, B., E. B. Rosen, E. Sideman, A. M. Shelton, and C. D. Smart. 2005. *Resource Guide for Organic Insect and Disease Management*. Ithaca: Cornell University Press.

Campbell, H., and R. Liepens. 2001. "Naming Organics: Understanding Organic Standards in New Zealand as a Discursive Field." *Sociologia Ruralis* 41, no. 1:21–39.

Canning, P. 2011. "A Revised and Expanded Food Dollar Series: A Better Understanding of Our Food Costs." U.S. Department of Agriculture, Economic Research Service. http://www.ers.usda.gov/publications/err -economic-research-report/err114.aspx. Accessed July 19, 2012.

Carfagna, Lindsey B., Emilie A. Dubois, Connor Fitzmaurice, Monique Y. Ouimette, Juliet B. Schor, Margaret Willis, and Thomas Laidley. 2014. "An Emerging Eco Habitus: The Reconfiguration of High Cultural Capital Practices among Ethical Consumers." *Journal of Consumer Culture* 14, no. 2:158–178.

Carson, R. 1994 [1962]. *Silent Spring*. Boston: Houghton Mifflin.

Charles, D. 2011. "Newbie Farmers Find That Dirt Isn't Cheap." *National Public Radio, Salt*. November 15. http://www.npr.org/blogs/thesalt/2011 /11/14/142305869/newbie-farmers-find-that-dirt-isnt-cheap. Accessed July 19, 2012.

Chiffoleau, Y. 2009. "From Politics to Co-operation: The Dynamics of Embeddedness in Alternative Food Supply Chains." *Sociologia Ruralis* 49, no. 3:218–235.

Cicatiello, C., B. Pancino, S. Pascucci, and S. Franco. 2015. "Relationship Patterns in Food Purchase: Observing Social Interactions in Different Shopping Environments." *Journal of Agricultural and Environmental Ethics* 28:21–42.

Claro, J. 2011. *Vermont Farmers Markets and Grocery Stores: A Price Comparison*. New England Organic Farming Association of Vermont.

Cloud, J. 2007. "Eating Better Than Organic." *Time Magazine*. http://www .time.com/time/magazine/article/0,9171,1595245,00.html. Accessed July 19, 2012.

Commoner, B. 1971. *The Closing Circle: Nature, Man, and Technology*, 1st ed. New York: Knopf.

Connolly, J., and A. Prothero. 2008. "Green Consumption: Life-Politics, Risk and Contradictions." *Journal of Consumer Culture* 8:117–145.

Constance, D. H., J. Y. Choi, and H. Lyke-Ho-Gland. 2008. "Conventionalization, Bifurcation, and Quality of Life: Certified and Non-Certified Organic Farms in Texas." *Southern Rural Sociology* 23, no. 1:208–234.

Coombes, B., and H. Campbell. 1998. "Dependent Reproduction of Alternative Modes of Agriculture: Organic Farming in New Zealand." *Sociologia Ruralis* 38, no. 2:127–145.

Cornucopia News. 2014. "Horizon 'Organic' Farm Accused of Improprieties, Again." *Cornucopia Institute*, February 14. http://www.cornucopia.org/2014/02/horizon-organic-factory-farm-accused-improprieties/. Accessed January 4, 2016.

Coslor, Erica. 2010. "Hostile Worlds and Questionable Speculation: Recognizing the Plurality of Views about Art and the Market." In *Economic Action in Theory and Practice: Anthropological Investigations* (Research in Economic Anthropology, vol. 30), ed. Donald Wood. Bingley, UK: Emerald.

Cowie, J., and N. Salvatore. 2008. "The Long Exception: Rethinking the Place of the New Deal in American History." *International Labor and Working-Class History* 74, no. 1:3–32.

DeLind, L. B. 1999. "Close Encounters with a CSA: The Reflections of a Bruised and Somewhat Wiser Anthropologist." *Agriculture and Human Values* 16:3–9.

———. 2000. "Transforming Organic Agriculture into Industrial Organic Products: Reconsidering National Organic Standards." *Human Organization* 59, no. 2:198–208.

———. 2003. "Considerably More Than Vegetables, a Lot Less Than Community: The Dilemma of Community Supported Agriculture." In *Fighting for the Farm*, ed. J. Adams, 192–206. Philadelphia: University of Pennsylvania Press.

———. 2011. "Are Local Food and the Local Food Movement Taking Us Where We Want to Go? Or Are We Hitching Our Wagons to the Wrong Stars?" *Agriculture and Human Values* 28:273–283.

DeMuth, S. 1993. "Defining Community Supported Agriculture." *Community Supported Agriculture (CSA): An Annotated Bibliography and Resource Guide*. U.S. Department of Agriculture. http://www.nal.usda.gov/afsic/pubs/csa/csadef.shtml. Accessed July 19, 2012.

DeVault, G. 2009. "The New USDA: A New Hope for Food?" *Mother Earth News*. http://www.motherearthnews.com/Sustainable Farming/USDA-Organic-Farms-Vilsack.aspx. Accessed July 18, 2012.

Dimitri, C., and C. Greene. 2002. "Recent Growth Patterns in the U.S. Organic Foods Market." U.S. Department of Agriculture, Economic Research Service, Agriculture Information Bulletin No. AIB-777.

Dimitri, C., and L. Oberholtzer. 2009. "Marketing U.S. Organic Foods: Recent Trends from Farms to Consumers." U.S. Department of Agriculture, Economic Research Service, Agriculture Information Bulletin No. 58.

Donahue, B. 1999. *Reclaiming the Commons: Community Farms and Forests in a New England Town.* New Haven: Yale University Press.

Drabenstott, M., and S. Moore. 2009. "Rural America in Deep Downturn." Kansas City, MO: Rural Policy Research Institute. http://www.rupri.org /Forms/CRC_Recession.pdf. Accessed July 19, 2012.

DuPuis, M., and S. Gillon. 2009. "Alternative Modes of Governance: Organic as Civic Engagement." *Agriculture and Human Values* 26:43–56.

DuPuis, M., and D. Goodman. 2005. "Should We Go 'Home' to Eat?: Toward a Reflexive Politics of Localism." *Journal of Rural Studies* 21:359–371.

Duram, L. A. 1997. "A Pragmatic Study of Conventional and Alternative Farmers in Colorado." *Professional Geographer* 49:202–213.

———. 1998. "Taking a Pragmatic Behavioral Approach to Alternative Agriculture Research." *American Journal of Alternative Agriculture* 13, no. 2:92–97.

———. 2000. "Agents' Perceptions of Structure: How Illinois Organic Farmers View Political, Economic, Social, and Ecological Factors." *Agriculture and Human Values* 17:35–48.

———. 2005. *Good Growing: Why Organic Farming Works.* Lincoln, NE: University of Nebraska Press.

Duram, L. A., and A. Mead. 2013. "Exploring Linkages between Consumer Food Co-operatives and Domestic Fair Trade in the United States." *Renewable Agriculture and Food Systems.* Available on CJO2013. doi:10.1017 /S1742170513000033.

Duscha, J. 1972. "Up, Up, Up—Butz Makes Hay Down on the Farm." *New York Times,* April 16.

Economic Research Service. 2014. "Number of U.S. Farmers Markets Continues to Rise." U.S. Department of Agriculture, Economic Research Service, August 4, 2014. http://ers.usda.gov. Accessed May 19, 2015.

———. 2015. "Farm Labor: Background." U.S. Department of Agriculture, Economic Research Service, October 20. http://ers.usda.gov. Accessed January 4, 2016.

Edleman, Lauren B., and Robin Stryker. 2005. "A Sociological Approach to Law and the Economy." In *Handbook of Economic Sociology,* 2nd ed., ed. Neil J. Smelser and Richard Swedberg, 527–551. Princeton, NJ: Princeton University Press, and New York: Russell Sage Foundation.

Edelman, Lauren B., Christopher Uggen, and Howard Erlanger. 1999. "The Endogeneity of Legal Regulation: Grievances Procedures as a Rational Myth." *American Journal of Sociology* 105:406–454.

Extension Toxicology Network. 1996. "Copper Sulfate." *Pesticide Information Profiles*. Corvallis, OR: Oregon State University.

Feber, R. E., L. G. Firbank, P. J. Johnson, and D. W. Macdonald. 1997. "The Effects of Organic Farming on Pest and Non-Pest Butterfly Abundance." *Agriculture Ecosystems and Environment* 64:133–139.

Feegan, R. B., and D. Morris. 2009. "Consumer Quest for Embeddedness: A Case Study of the Brandtford Farmers' Market." *International Journal of Consumer Studies* 33:235–243.

Feenstra, G. 2002. "Creating Space for Sustainable Food Systems: Lessons from the Field." *Agriculture and Human Values* 19:99–106.

Fligstein, Neil. 2005. "The Political and Economic Sociology of International Economic Arrangements." In *Handbook of Economic Sociology*, 2nd ed., ed. Neil J. Smelser and Richard Swedberg. Princeton, NJ: Princeton University Press, and New York: Russell Sage Foundation.

Follett, J. R. 2009. "Choosing a Food Future: Differentiating among Alternative Food Options." *Journal of Agricultural and Environmental Ethics* 22:31–51.

Food and Agriculture Organization. 2010. "Crop Biodiversity: Use It or Lose It." U.N. Food and Agriculture Organization. http://www.fao.org/news/story/en/item/46803/icode/. Accessed July 19, 2012.

Foster, J. B. 1999. *The Vulnerable Planet: A Short Economic History of the Environment*. New York: Monthly Review Press.

———. 2002. *Ecology against Capitalism*. New York: Monthly Review Press.

Francis, C. A. 2009. *Organic Farming: The Ecological System*. Madison, WI: American Society of Agronomy, Crop Science Society of America, Soil Science Society of America.

Freyfogle, E. T. 2001. "Introduction: A Durable Scale." In *The New Agrarianism: Land, Culture, and the Community of Life*, ed. E. T. Freyfogle. Washington, D.C.: Island Press.

Fromartz, S. 2006. *Organic Inc.: Natural Foods and How They Grew*. Orlando, FL: Harcourt.

Gareau, B. J. 2008. "Dangerous Holes in Global Environmental Governance: The Roles of Neo-liberal Discourse, Science, and California Agriculture in the Montreal Protocol." *Antipode* 40, no. 1:102–130.

———. 2013. *From Precaution to Profit: Contemporary Challenges to Environmental Protection in the Montreal Protocol*. Yale Agrarian Studies Series. New Haven: Yale University Press.

Gareau, B. J., and J. Borrego. 2012. "Global Environmental Governance, Competition, and Sustainability in Global Agriculture." In *Handbook of World-Systems Analysis*, ed. S. Babones and C. Chase-Dunn, 357–365. New York: Routledge.

Gilbert, Jess, Gwen Sharp, and Sindy M. Felin. 2002. "The Loss and Persistence of Black-Owned Farms and Farmland: A Review of the Research Literature and Its Implications." *Southern Rural Sociology* 18:1–30.

Gillespie, Gilbert, Duncan L. Hilchey, C. Clare Hinrichs, and Gail Feenstra. 2007. "Farmers' Markets as Keystones in Rebuilding Local and Regional Food Systems." In *Rebuilding the North American Food System: Strategies for Sustainability*, ed. C. Clare Hinrichs and Thomas A. Lyson, 65–84. Lincoln, NE: University of Nebraska Press.

Gliessman, S. R. 2006. *Agroecology: The Ecology of Sustainable Food Systems*, 2nd ed. Boca Raton, FL: CRC Press.

Goffman, E. 1989. "On Fieldwork." *Journal of Contemporary Ethnography* 18:123–132.

Gold, M. V. 2007. "Organic Production/Organic Food: Information Access Tools." U.S. Department of Agriculture, Alternative Farming Systems Information Center. http://www.nal.usda.gov/afsic/pubs/ofp/ofp.shtml. Accessed July 17, 2012.

Goodman, D., B. Sorj, and J. Wilkenson. 1987. *From Farming to Biotechnology*. Oxford: Blackwell.

Gourevitch, P. 2011. "The Value of Ethics: Monitoring Normative Compliance in Ethical Consumption Markets." In *The Worth of Goods: Valuation and Pricing in the Economy*, ed. J. Beckert and P. Aspers, 86–105. Oxford: Oxford University Press.

Gowan, T., and R. Slocum. 2014. "Artisanal Production, Communal Provisioning, and Anticapitalist Politics in the Aude, France." In Schorand Thompson, *Sustainable Lifestyles and the Quest for Plentitude*, 27–62.

Granovetter, M. 1985. "Economic Action and Social Structure: The Problem of Embeddedness." *American Journal of Sociology* 91, no. 3:481–510.

Grasseni, C. 2003. "Packaging Skills: Calibrating Cheese to the Global Market." In *Commodifying Everything*, ed. S. Strasser. New York: Routledge.

———. 2011. "Re-inventing Food: Alpine Cheese in the Age of Global Heritage." *Anthropology of Food*. http://aof.revues.org/6819. Accessed May 19, 2015.

———. 2014. "Of Cheese and Ecomuseums: Food as Cultural Heritage in the Northern Italian Alps." In *Edible Identities: Food as Cultural Heritage*, ed. M. Di Giovine and R. L. Brulotte, 55–66. Burlington, VT: Ashgate.

Gray, M. 2013. *Labor and the Locavore: The Making of a Comprehensive Food Ethic*. Berkeley, CA: University of California Press.

Greene, C., and A. Kremen. 2003. *U.S. Organic Farming in 2000–2001*. Agriculture Information Bulletin 780. Washington, D.C.: U.S. Department of Agriculture, Economic Research Service.

Greene, W. 1971. "Guru of the Organic Cult." *New York Times*, June 6.

Grover, J., and M. Goldberg. 2010. "False Claims, Lies Caught on Tape at Farmers Markets." *NBC Southern California*, September 23. http://www .nbclosangeles.com/news/local/Hidden-Camera-Investigation-Farmers -Markets-103577594.html. Accessed July 19, 2012.

Guthman, J. 1998. "Regulating Meaning: The Codification of California Organic Agriculture." *Antipode* 30, no. 2:135–154.

———. 2003. "Fast Food/Organic Food: Reflexive Tastes and the Making of Yuppie Chow." *Social and Cultural Geography* 4, no. 1:45–58.

———. 2004a. *Agrarian Dreams*. Berkeley, CA: University of California Press.

———. 2004b. "Back to the Land: The Paradox of Organic Food Standards." *Environment and Planning* 36:511–528.

———. 2004c. "The Trouble with 'Organic Lite' in California: A Rejoinder to the 'Conventionalization' Debate." *Sociologia Ruralis* 44, no. 3:301–316.

———. 2007. "The Polanyian Way? Voluntary Food Labels and Neoliberal Governance." *Antipode* 39, no. 3:456–478.

———. 2008a. "'If They Only Knew': Color Blindness and Universalism in California's Alternative Food Institutions." *Professional Geographer* 60, no. 3:387–397.

———. 2008b. "Neoliberalism and the Making of Food Politics in California." *Geoforum* 39, no. 3:1171–1183.

———. 2011. *Weighing In: Obesity, Food Justice, and the Limits of Capitalism*. Berkeley: University of California Press.

Halweil, B. 2004. *Eat Here: Reclaiming Homegrown Pleasures in a Global Supermarket*. Washington, D.C.: Worldwatch Institute.

Haney, D. Q. 1984. "Doctors Trace Drug-Resistant Germs from Cattle to People." *Associated Press*, September 5.

Hansen, L. 2011. "America's Future Farmers Already Dropping Away." Interview with Secretary of Agriculture Tom Vilsack. National Public

Radio, February 27. http://www.npr.org/2011/02/27/134103432/Americas
-Future-Farmers-Already-Dropping-Away. Accessed July 19, 2012.

Harrison, J. L. 2011. *Pesticide Drift and the Pursuit of Environmental Justice.*
Cambridge, MA: MIT Press.

Hassanein, N. 1999. *Changing the Way America Farms: Knowledge and Community in the Sustainable Agriculture Movement.* Lincoln, NE: University
of Nebraska Press.

Healy, Kieran. 2006. *Last Best Gifts: Altruism and the Market for Human Blood
and Organs.* Chicago: University of Chicago Press.

Hennessy, M. 2013. "WhiteWave Foods Enters the Produce Aisle with
Earthbound Organic Acquisition." *Food Navigator-USA*, December 10.
http://www.foodnavigator-usa.com/Manufacturers/WhiteWave-Foods
-enters-produce-aisle-with-Earthbound-Organic-acquisition. Accessed
January 4, 2016.

Hertz, T. 2014. "Farm Labor." U.S. Department of Agriculture, Economic
Research Service. http://www.ers.usda.gov/topics/farm-economy/farm
-labor.aspx. Accessed December 27, 2015.

Hess, D. J. 2009. *Localist Movements in a Global Economy: Sustainability, Justice, and Urban Development in the United States.* Cambridge, MA: MIT
Press.

Heynen, Nik. 2009. "Bending the Bars of Empire from Every Ghetto for
Survival: The Black Panther Party's Radical Antihunger Politics of Social Reproduction and Scale." *Annals of the Association of American Geog-
raphers* 99, no. 2:406–422.

Hinrichs, C. C. 2000. "Embeddedness and Local Food Systems: Notes
on Two Types of Direct Agriculture Market." *Journal of Rural Studies*
16:295–303.

———. 2010. "Conceptualizing and Creating Sustainable Food Systems: How Interdisciplinarity Can Help." In *Imagining Sustainable Food
Systems Theory and Practice*, ed. Alison Blay-Palmer, 17–36. Farnham, Sur-
rey: Ashgate.

Hinrichs, C. C., and P. Allen. 2008. "Selective Patronage and Social Justice: Local Food Consumer Campaigns in Historical Context." *Journal
of Agricultural and Environmental Ethics* 21:329–352.

Ho, K. 2009. *Liquidated: An Ethnography of Wall Street.* Durham, NC: Duke
University Press.

Hoang, Kimberly Kay. 2011. "'She's Not a Low-Class Dirty Girl!': Sex
Work in Ho Chi Minh City." *Journal of Contemporary Ethnography* 40,
no. 4:367–396.

———. 2015. *Dealing in Desire: Asian Ascendency, Western Decline, and the Hidden Currencies of Global Sex Work*. Berkeley, CA: University of California Press.

Holmes, S. 2007. "'Oaxacans Like to Work Bent Over': The Naturalization of Social Suffering among Berry Farm Workers." *International Migration* 45, no. 3:39–68.

———. 2013. *Fresh Fruit, Broken Bodies: Migrant Farmworkers in the United States*. Berkeley, CA: University of California Press.

Horovitz, B. 2006. "More University Students Call for Organic, 'Sustainable' Food; Campuses Nationwide Buy More Food from 'Local' Farms." *USA Today*, September 27.

Howard, P. H. 2009. "Consolidation in the North American Organic Food Processing Sector: 1997–2007." *International Journal of Sociology of Agriculture and Food* 16, no. 1:13–30.

———. 2015. "Organic Processing Industry Structure: Acquisitions & Alliances, Top 100 Food Processors in North America," December. https://msu.edu/~howardp/organicindustry.html. Accessed January 4, 2016.

Howard, P. H., and P. Allen. 2008. "Consumer Willingness to Pay for Domestic 'Fair Trade': Evidence from the United States." *Renewable Agriculture and Food Systems* 23:235–242.

———. 2010. "Beyond Organic and Fair Trade? An Analysis of Ecolabel Preferences in the United States." *Rural Sociology* 75, no. 2:244–269.

IFOAM. 2013. "The Principles of Organic Agriculture." International Federation of Organic Agriculture Movements. http://infohub.ifoam.org/en/what-organic/principles-organic-agriculture. Accessed July 26, 2013.

Ikerd, J. 2001. "The Architecture of Organic Production." Inaugural National Organics Conference, Sydney, Australia. August 27–28. http://web.missouri.edu/~ikerdj/papers/Australia.html. Accessed July 19, 2012.

Inhetveen, Heide. 1998. "Women Pioneers in Farming: A Gendered History of Agricultural Progress." *Sociologia Ruralis* 38, no. 3:265–284.

Jacobson, M. F. 1972. "Feeding the People, Not Food Producers." *New York Times*, August 31.

Jaffee, D., and P. H. Howard. 2010. "Corporate Cooptation of Organic and Fair Trade Standards." *Agriculture and Human Values* 27:387–399.

Jager, R. 2004. *The Fate of Family Farming: Variations on an American Idea*. Lebanon, NH: University Press of New England.

Johnston, J. 2008. "The Citizen-Consumer Hybrid: Ideological Tensions and the Case of Whole Foods Market." *Theory and Society* 37:229–270.

Johnston, J., and S. Baumann. 2010. *Foodies: Democracy and Distinction in the Gourmet Foodscape.* New York: Routledge.

Johnston, J., and M. Szabo. 2011. "Reflexivity and the Whole Foods Market Consumer: The Lived Experience of Shopping for Change." *Agriculture and Human Values* 28:303–319.

Johnston, J., A. Biro, and N. MacKendrick. 2009. "Lost in the Supermarket: The Corporate-Organic Foodscape and the Struggle for Food Democracy." *Antipode* 41, no. 3:509–532.

Jonsson, P. 2006. "A Comeback for Small Farms." *Christian Science Monitor,* February 9. http://www.csmonitor.com/2006/0209/p03s03ussc.html. Accessed July 19, 2012.

Josselson, R. 2011. "Narrative Research: Constructing, Deconstructing, and Reconstructing Story." In *Five Ways of Doing Qualitative Analysis,* ed. Frederick J. Wertz et al. New York: Guilford Press.

Kautsky, Karl. 1988 [1899]. *The Agrarian Question.* London: Zwan Publications.

Keough, G. 2014. "Massachusetts Agriculture Defies Trends." U.S. Department of Agriculture blog, July 7. http://blogs.usda.gov/2014/07/07/massachusetts-agriculture-defies-national-trends/. Accessed August 18, 2014.

Khan, S. 2012. *Privilege: The Making of an Adolescent Elite at St. Paul's School.* Princeton, NJ: Princeton University Press.

Kirschenmann, F. 2004. "Ecological Morality: A New Ethic for Agriculture." In *Agroecosystems Analysis,* ed. D. Rickerl and C. Francis, 167–176. Madison, WI: American Society of Agronomy.

Kohlhepp, J. 2011. "CSA: The Sustenance of Small Farms." *Tri-Town News,* August 25.

Krippner, Greta R. 2001. "The Elusive Market: Embeddedness and the Paradigm of Economic Sociology." *Theory and Society* 30, no. 6:775–810.

Kristiansen, P., A. Taji, and J. Reganold, eds. 2006. *Organic Agriculture: A Global Perspective.* Ithaca, NY: Cornell University Press.

Kummer, C. 2010. "The Great Grocery Smackdown: Will Walmart, Not Whole Foods, Save the Small Farm and Make America Healthy?" *Atlantic.* http://www.theatlantic.com/magazine/archive/2010/03/the-great-grocery-smackdown/7904/. Accessed July 22, 2012.

Kusenbach, M. 2003. "Street Phenomenology: The Go-Along as Ethnographic Research Tool." *Ethnography* 4:455–485.

Lachman, G. 2007. *Rudolf Steiner: An Introduction to His Life and Work.* New York: Penguin.

La Gorce, T. 2011. "Organic Blueberries Don't Come Easily." *New York Times*, June 17.

Läpple, D. 2012. "Comparing Attitudes and Characteristics of Organic, Former Organic and Conventional Farmers: Evidence from Ireland." *Renewable Agriculture and Food Systems*. Available on CJO2012. doi:10.1017/S1742170512000294.

League of Conservation Voters. 1983. "How Congress Voted on Energy and the Environment: 1982 Voting Chart." Washington, D.C.: League of Conservation Voters.

Leary, W. E. 1989. "Ideas and Trends: Fear of Aflatoxin; The Debate about the Carcinogens That Man Didn't Make." *New York Times*, March 5.

Lichtenstein, N. 2009. *The Retail Revolution: How Wal-Mart Created a Brave New World of Business.* New York: Metropolitan Books.

Lin, B. H., T. A. Smith, and C. L. Huang. 2008. "Organic Premiums of U.S. Fresh Produce." Renewable Agriculture and Food Systems 23:208–216.

Linebaugh, P. 2008. *The Magna Carta Manifesto: Liberties and Commons for All.* Berkeley CA: University of California Press.

Lipson, M. 1997. *Searching For the 'O-Word': Analyzing the USDA Current Research Information System for Pertinence to Organic Farming.* Santa Cruz, CA: Organic Farming Research Foundation.

Local Harvest. 2015. "Community Supported Agriculture." http://www.localharvest.org/csa/. Accessed January 4, 2016.

Lockeretz, W. 1997. "Diversity of Personal and Enterprise Characteristics among Organic Growers in the Northeastern United States." *Biological Agriculture and Horticulture* 14:13–24.

Lockie, S., K. Lyons, G. Lawrence, and J. Grice. 2004. "Choosing Organics: A Path Analysis of Factors Underlying the Selection of Organic Food among Australian Consumers." *Appetite* 43:135–146.

Lockie, S., K. Lyons, G. Lawrence, and D. Halpin. 2006. *Going Organic: Mobilizing Networks for Environmentally Responsible Food Production.* Wallingford, Oxfordshire, UK: CAB International.

Lockie, S., K. Lyons, G. Lawrence, and K. Mummery. 2002. "Eating 'Green': Motivations behind Organic Food Consumption in Australia." *Sociologia Ruralis* 42, no. 1:23–40.

Lohr, L. 2009. "1990 Farm Bill, Title XXI." In *Encyclopedia of Organic, Local, and Sustainable Food*, ed. Leslie Duram. Santa Barbara, CA: Greenwood.

Lyons, K., and G. Lawrence. 2001. "Institutionalisation and Resistance: Organic Agriculture in Australia and New Zealand." In *Food, Nature and Society: Rural Life in Late Modernity*, ed. H. Tovey and M. Blanc. Ashgate: Aldershot.

Lyson, T. A., and J. Green. 1999. "The Agricultural Marketscape: A Framework for Sustaining Agriculture and Communities in the Northeast." *Journal of Sustainable Agriculture* 15, nos. 2–3:133–150.

Macaulay, Stewart. 1963. "Non-Contractual Relations in Business: A Preliminary Study." *American Sociological Review* 28, no. 1:1–19.

MacKenzie, D., and Y. Millo. 2003. "Constructing a Market, Performing Theory: The Historical Sociology of a Financial Derivatives Exchange." *American Journal of Sociology* 109:107–145.

Magdoff, F., J. B. Foster, and F. H. Buttel, eds. 2000. *Hungry for Profit: The Agribusiness Threat to Farmers, Food, and the Environment*. New York: Monthly Review Press.

Major, W. H. 2011. *Grounded Vision: New Agrarianism and the Academy*. Tuscaloosa, AL: University of Alabama Press.

Marcuse, H. 1964. *One-Dimensional Man*. Boston: Beacon Press.

Mariola, M. J. 2008. "The Local Industrial Complex? Questioning the Link between Local Foods and Energy Use." *Agriculture and Human Values* 25:193–196.

Martin, A., and K. Severson. 2008. "Sticker Shock in the Organic Aisle." *New York Times*, April 18.

Marx, Karl. 1977 [1867]. *Capital*, vol. 1. Translated by Ben Fowkes. New York: Vintage.

Marx de Salcedo, A. 2007. "The Bunny v. the Blue Box." *Salon*, January 30. http://www.salon.com/2007/01/30/annies/. Accessed January 4, 2016.

Mascarenhas, M., and L. Busch. 2006. "Seeds of Change: Intellectual Property Rights, Genetically Modified Soybeans and Seed Saving in the United States." *Sociologia Ruralis* 46, no. 2:122–138.

Mauss, Marcel. 1954. *The Gift: Forms and Functions of Exchange in Archaic Societies*. Glencoe, IL: Free Press.

McClain, N., and A. Mears. 2012. "Free to Those Who Can Afford It: The Everyday Affordance of Privilege." *Poetics* 40:133–149.

McEntee, J. C. 2011. "Realizing Food Justice: Divergent Locals in the Northeastern United States." In *Cultivating Food Justice: Race, Class, and Sustainability*, ed. A. H. Alkon and J. Agyeman. Cambridge, MA: MIT Press.

McKibben, B. 2007. *Deep Economy*. New York: Henry Holt.

McMichael, P. 2005. "Global Development and the Corporate Food Regime." *Research in Rural Sociology and Development* 11:265–299.

————. 2010. *Contesting Development: Critical Struggles for Social Change.* New York: Routledge.

McPherson, M., L. Smith-Lovin, and M. E. Brashears. 2006. "Social Isolation in America: Changes in Core Discussion Networks of Two Decades." *American Sociological Review* 71:353–375.

Mears, A. 2011. "Pricing Looks: Circuits of Value in Fashion Modeling Markets." In *The Worth of Goods: Valuation and Pricing in the Economy,* ed. J. Beckert and P. Aspers, 155–177. Oxford: Oxford University Press.

Mills, C. W. 2000 [1959]. *The Sociological Imagination.* Oxford: Oxford University Press.

Moore, D., A. Pandian, and J. Kosek. 2003. *Race, Nature, and the Politics of Difference.* Durham, NC: Duke University Press.

Moskin, J. 2009. "Northeast Tomatoes Lost, and Potatoes May Follow." *New York Times,* July 28. http://www.nytimes.com/2009/07/29/dining/29toma.html. Accessed July 19, 2012.

Mount, P. 2012. "Growing Local Food: Scale and Local Food Systems Governance." *Agriculture and Human Values* 29, no. 1:107–121.

Munoz, O. 2010. "More Farmers Work Away from Fields to Pay Bills." *Bismarck Tribune,* December 10.

Murphy, K. 1996. "Organic Food Makers Reap Green Yields of Revenue: A Widening Popularity Brings Acquisitions." *New York Times,* October 26.

National Academy of Sciences. 1975. *Annual Report: Fiscal Years 1973 and 1974.* Washington D.C.: National Academy of Sciences.

National Agricultural Statistics Service. 2011. *Land Values 2011 Summary.* U.S. Department of Agriculture. http://usda01.library.cornell.edu/usda/current/AgriLandVa/AgriLandVa-08-04-2011.pdf. Accessed July 19, 2012.

National Park Service. "Trades along the Battle Road." Marker at Minute Man National Historic Park.

Ness, C. 2006. "Whole Foods, Taking Flak, Thinks Local." *San Francisco Chronicle,* July 26.

Netting, R. M. 1993. *Smallholders, Householders: Farm Families and the Ecology of Intensive, Sustainable Agriculture.* Stanford, CA: Stanford University Press.

Nettle, C. 2014. *Community Gardening as Social Action.* Farnham: Ashgate.

Nichols, J. 2003. "Needed: A Rural Strategy." *Nation,* November 3. http://www.thenation.com/article/needed-rural-strategy#. Accessed July 19, 2012.

Oakes, J. B. 1989. "A Silent Spring, for Kids." *New York Times,* March 30.

Obach, B. K. 2007. "Theoretical Interpretations of the Growth in Organic Agriculture: Agricultural Modernization or an Organic Treadmill?" *Society and Natural Resources* 20, no. 3:229–244.

Oberholtzer, L. 2009. "'Direct Marketing' and 'Suburban Sprawl.'" In *Encyclopedia of Organic, Local, and Sustainable Food*, ed. Leslie Duram. Santa Barbara, CA: Greenwood.

Oberholtzer, L., K. Clancy, and J. D. Esseks. 2010. "The Future of Farming on the Urban Edge: Insights from Fifteen U.S. Counties about Farmland Protection and Farm Viability." *Journal of Agriculture, Food Systems, and Community Development* 1, no. 2:59–75.

Oldenberg, R. 1989. *The Great Good Place*. New York: Paragon House.

O'Neill, J. M. 2009. "What's Killing Our Tomatoes? Late-Blight Fungus Ruining Crops in 13 States." *The Record*, July 18.

OFPA. (Organic Foods Production Act). 1990. *Code of Federal Regulations*.

OTA (Organic Trade Association). 2005. "A National Organic Initiative." http://www.ota.com/pics/documents/NationalOrganicInitiative_206 .pdf. Accessed July 18, 2012.

———. 2015. "There's More to Organic than Meets the Eye." https://www .ota.com/resources/market-analysis. Accessed December 2, 2015.

Pechlaner, Gabriela, and Gerardo Otero. 2008. "The Third Food Regime: Neoliberal Globalism and Agricultural Biotechnology in North America." *Sociologia Ruralis* 48, no. 4:351–371.

Petersen, C. 1974. "The 'Anarchlings' Who Bark but Never Bite." *Chicago Tribune*, May 24.

Pilgeram, R. 2011. "'The Only Thing That Isn't Sustainable . . . Is the Farmer': Social Sustainability and the Politics of Class among Pacific Northwest Farmers Engaged in Sustainable Farming." *Rural Sociology* 76, no. 3:375–393.

Pirog, R., and N. McCann. 2009. *Is Local Food More Expensive? A Consumer Price Perspective on Local and Non-Local Foods Purchased in Iowa*. Ames, IA: Leopold Center.

Pisani Gareau, T., N. DeBarros, M. Barbercheck, and D. Mortensen. 2009. *Conserving Wild Bees in Pennsylvania*. CODE# UF023 R3M04/10payne5005. State College, PA: Pennsylvania State University, Agricultural Communications and Marketing.

Polanyi, K. 1957. *The Great Transformation: The Political and Economic Origins of Our Time*. Boston: Beacon Press.

Pollan, M. 2006. *The Omnivore's Dilemma: A Natural History of Four Meals*. New York: Penguin.

Pratt, Jeffery. 2008. "Food Values: The Local and the Authentic." *Hidden Hands in the Market: Ethnographies of Fair Trade, Ethical Consumption, and Corporate Responsibility*. Special issue of *Research in Economic Anthropology* 28:53–70.

———. 2009. "Incorporation and Resistance: Analytical Issues in the Conventionalization Debate and Alternative Food Chains." *Journal of Agrarian Change* 9, no. 2:155–174.

Pretty, J. 1998. *The Living Land: Agriculture, Food, and Community Regeneration in Rural Europe*. London: Earthscan Publications.

Pritchard, B., D. Burch, and G. Lawrence. 2007. "Neither 'Family' nor 'Corporate' Farming: Australian Tomato Growers as Farm Family." *Journal of Rural Studies* 23:75–87.

Putnam, R. 2000. *Bowling Alone: The Collapse and Revival of American Community*. New York: Simon and Schuster.

Raftery, I. 2011. "Young Farmers Find Huge Obstacles to Getting Started." *New York Times*, November 12.

Ramde, D. 2011. "More Young People See Opportunities in Farming." *USA Today*, December 22. http://www.usatoday.com/money/industries/food/story/2011-12-24/young-people-farming/52163914/1. Accessed July 19, 2012.

Reed, Matthew. 2001. "Fight the Future! How the Contemporary Campaigns of the UK Organic Movement Have Arisen from Their Composting of the Past." *Sociologia Ruralis* 41, no. 1:131–145.

Rosenbaum, D. E. 1984. "States' Actions of EDB in Food Resulting in Pattern of Confusion." *New York Times*, February 18.

Rosin, C., and H. Campbell. 2009. "Beyond Bifurcation: Examining the Conventions of Organic Agriculture in New Zealand." *Journal of Rural Studies* 25:35–47.

Russell, H. S. 1976. *A Long, Deep Furrow: Three Centuries of Farming in New England*. Lebanon, NH: University Press of New England.

Sandilands, C. 1993. "On 'Green' Consumerism: Environmental Privatization and 'Family Values.'" *Canadian Women's Studies* 13, no. 3:45–47.

Saunders, C., A. Barber, and G. Taylor. 2006. "Food Miles—Comparative Energy/Emissions Performance of New Zealand's Agricultural Industry." Christchurch, New Zealand: Lincoln University. Report No. 285.

Sayre, L. 2011. "The Politics of Organic Farming: Populists, Evangelicals, and the Agriculture of the Middle." *Gastronomic: The Journal of Food and Culture* 11, no. 2:38–47.

Schor, J. B. 1991. *The Overworked American: The Unexpected Decline of Leisure*. New York: Basic Books.

———. 1996. "Summary of 'The Insidious Cycle of Work and Spend.'" In *The Consumer Society*, ed. N. R. Goodwin, F. Ackerman, and D. Kiron. Washington, D.C.: Island Press.

———. 1998. *The Overspent American: Upscaling, Downshifting, and the New Consumer.* New York: Basic Books.

———. 2010. *Plentitude: The New Economics of Wealth.* New York: Penguin.

Schor, J. B., and C. Fitzmaurice. 2015. "Sharing, Collaborating, and Connecting: The Emergence of the Sharing Economy." In *Handbook of Research on Sustainable Consumption*, ed. Lucia A. Reisch, and John Thøgersen. Northhampton, MA: Edward Elgar.

Schor, J. B., and C. J. Thompson, eds. 2014. *Sustainable Lifestyles and the Quest for Plentitude: Case Studies of the New Economy.* New Haven: Yale University Press.

Schurman, R., and W. Munro. 2009. "Targeting Capital: A Cultural Economy Approach to Understanding the Efficacy of Two Anti-Genetic Engineering Movements." *American Journal of Sociology* 115, no. 1:155–202.

Seyfang, G. 2006. "Ecological Citizenship and Sustainable Consumption: Examining Local Organic Food Networks." *Journal of Rural Studies* 22:383–395.

Shabecoff, P. 1983. "Florida's Ban on 26 Food Products Prompts EPA Pesticide Investigation." *New York Times*, December 22.

Sierra, L., K. Klonsky, R. Strochlic, S. Brody, and R. Molinar. 2008. *Factors Associated with Deregistration among Organic Farmers in California.* Davis, CA: Sustainable Agriculture Research and Education Program.

Silverman, G. 2011. "Local Farmers Question the Economic Benefit of Organic Label." *Medill Reports*, May 26. Chicago: Northwestern University. http://news.medill.northwestern.edu/chicago/news.aspx?id=186782. Accessed July 19, 2012.

Small, M. 2009. *Unanticipated Gains: Origins of Network Inequality in Everyday Life.* Oxford: Oxford University Press.

Soil Association Certification. 2013. http://www.sacert.org/. Accessed March 4, 2015.

Stock, P. V. 2007. "'Good Farmers' as Reflexive Producers: An Examination of Family Organic Farmers in the U.S. Midwest." *Sociologia Ruralis* 47, no. 2:83–102.

Swedberg, Richard. 2003. "The Case for an Economic Sociology of Law." *Theory and Society* 32:1–37.

Thirsk, J. 1997. *Alternative Agriculture: A History from the Black Death to the Present Day.* Oxford: Oxford University Press.

Thompson, C., and G. Coskuner-Balli. 2007a. "Countervailing Market Responses to Corporate Co-optation and the Ideological Recruitment of Consumption Communities." *Journal of Consumer Research* 34, no. 2: 135–152.

———. 2007b. "Enchanting Ethical Consumerism: The Case of Community Supported Agriculture." *Journal of Consumer Culture* 7:275–303.

Thompson, C., and M. Press. 2014. "How Community-Supported Agriculture Facilitates Reembedding and Reterritorializing Practices of Sustainable Consumption." In Schor and Thompson, *Sustainable Lifestyles and the Quest for Plentitude*, 125–147.

Trauger, A., C. Sachs, M. Barbercheck, K. Brasier, and N. E. Kiernan. 2010. "'Our Market Is Our Community': Women Farmers and Civic Agriculture in Pennsylvania, USA." *Agriculture and Human Values* 27:43–55.

UMass Extension. 2012. "Tomato and Potato Late Blight." University of Massachusetts, Amherst, Center for Agriculture. http://extension.umass .edu/vegetable/diseases/tomato-late-blight. Accessed July 19, 2012.

Upright, C. B. 2012. "New-Wave Cooperatives Selling Organic Food: The Curious Endurance of an Organizational Form." PhD dissertation, Department of Sociology, Princeton University. Available at http://arks .princeton.edu/ark:/88435/dsp012f75r8052.

U.S. Congress. 1982. "Organic Farming Act of 1982: Hearing before the Subcommittee on Forests." H.R. 5618, 97th Congress, 2nd Session. http:// www.archive.org/stream/organicfarmingaooenergoog#page/no/mode /2up. Accessed July 18, 2012.

USDA. 1980. "Report and Recommendations on Organic Farming." http:// www.nal.usda.gov/afsic/pubs/USDAOrgFarmRpt.pdf. Accessed July 18, 2012.

———. 2009. "2007 Census of Agriculture: Farm Numbers." *2007 Census of Agriculture.* http://www.agcensus.usda.gov/Publications/2007/Online _Highlights/Fact_Sheets/Farm_Numbers/farm_numbers.pdf. Accessed July 19, 2012.

———. 2014. "2012 Census Publications." *2012 Census of Agriculture.* http:// www.agcensus.usda.gov/Publications/2012/. Accessed May 19, 2015.

Vanac, Mary. 2013. "Kathleen Merrigan Abruptly Resigns." *Columbus Dispatch*, March 15. Available at http://www.dispatch.com/content/blogs/the -bottom-line/2013/03/usda-deputy-secretary-kathleen-merrigan -resigns.html. Accessed December 15, 2015.

Veblen, Thorstein. 1912 [1899]. *The Theory of the Leisure Class.* New York: MacMillan.

Velthius, O. 2004. "An Interpretive Approach to the Meaning of Prices." *Review of Austrian Economics* 17, no. 4:371–386.

Venkataraman, B. 2009. "Late Blight Yields Bitter Harvest." *Boston Globe*, July 31. http://www.boston.com/news/local/massachusetts/articles/2009 /07/31/disease_that_spawned_irelands_potato_famine_hits_new _england/?page=2. Accessed July 19, 2012.

Wacquant, L. 2004. *Body and Soul: Notebooks of an Apprentice Boxer.* Oxford: Oxford University Press.

Walton, J. 1992. "Making the Theoretical Case." In *What Is a Case? Exploring the Foundations of Social Inquiry*, ed. C. Ragin and H. Becker. Cambridge: Cambridge University Press.

Walz, E. 2004. "Final Results of the Fourth National Organic Farmers' Survey: Sustaining Organic Farms in a Changing Organic Marketplace." Santa Cruz, CA: Organic Farming Research Foundation. http:// ofrf.org/publications/pubs/4thsurvey_results.pdf. Accessed July 19, 2012.

Warner, M. 2005. "What Is Organic? Powerful Players Want a Say." *New York Times*, November 1.

———. 2006. "Wal-Mart Eyes Organic Foods." *New York Times*, May 12.

Weber, J., and M. Ahearn. 2012. "Farm Household Well-Being: Labor Allocations and Age." U.S. Department of Agriculture, Economic Research Service. http://www.ers.usda.gov/topics/farm-economy/farm-household -well-being/labor-allocations-age.aspx. Accessed July 19, 2012.

Weise, E. 2009. "On Tiny Plots, A New Generation of Farmers Emerges." *USA Today*, July 14. http://www.usatoday.com/news/nation/environment /2009-07-13-young-farmers_N.htm. Accessed July 19, 2012.

Weiss, Brad. 2011. "Making Pigs Local: Discerning the Sensory Character of Place." *Cultural Anthropology* 26, no. 3:438–461.

Wherry, F. F. 2012. "Performance Circuits in the Marketplace." *Politics and Society* 40, no. 2:203–221.

Winders, B. 2009. *The Politics of Food Supply: U.S. Agricultural Policy in the World Economy.* New Haven: Yale University Press.

Wood, Spencer D., and Jess Gilbert. 2000. "Returning African American Farmers to the Land: Recent Trends and a Policy Rationale." *Review of Black Political Economy* 27, no. 4:43–64.

Young, K. 2015. "Sales from U.S. Organic Farms Up 72 Percent, USDA Reports." U.S. Department of Agriculture, Census of Agriculture, September 17. http://www.agcensus.usda.gov/Newsroom/2015/09_17_2015.php. Accessed January 4, 2016.

Youngberg, G., and S. P. DeMuth. 2013. "Organic Agriculture in the United States: A 30 Year Retrospective." *Renewable Agriculture and Food Systems* 28, no. 4:294–328.

Zelizer, V. A. 1988. "Beyond the Polemics on the Market: Establishing a Theoretical and Empirical Agenda." *Sociological Forum* 3, no. 4:614–634.

———. 2005. *The Purchase of Intimacy.* Princeton, NJ: Princeton University Press.

———. 2006. "Money, Power, and Sex." *Yale Journal of Law and Feminism* 18:303–315.

———. 2007. "Pasts and Futures of Economic Sociology." *American Behavioral Scientist* 50, no. 8:1056–1069.

———. 2010. *Economic Lives: How Culture Shapes the Economy.* Princeton, NJ: Princeton University Press.

———. 2012. "How I Became a Relational Economic Sociologist and What Does That Mean?" *Politics and Society* 40, no. 2:145–174.

INDEX

Numbers in *italics* indicate figures.